CW01379254

PURITY AND POLLUTION

STUDIES IN GENDER HISTORY

By challenging long-accepted approaches, categories and priorities, gender history has necessitated nothing less than a change in the historical terrain. This series seeks to publish the latest and best research, which not only continues to restore women to history and history to women, but also to encourage the development of a new channel of scholarship.

Published titles include:

Kathryn Gleadle
THE EARLY FEMINISTS

Elizabeth C. Sanderson
WOMEN AND WORK IN EIGHTEENTH-CENTURY EDINBURGH

Pamela Sharpe
ADAPTING TO CAPITALISM

Lilian Lewis Shiman
WOMEN AND LEADERSHIP IN NINETEENTH-CENTURY ENGLAND

Clare Taylor
WOMEN OF THE ANTI-SLAVERY MOVEMENT

Purity and Pollution

Gender, Embodiment and Victorian Medicine

Alison Bashford
Lecturer in Women's Studies and History
University of Sydney
Australia

First published in Great Britain 1998 by
MACMILLAN PRESS LTD
Houndmills, Basingstoke, Hampshire RG21 6XS and London
Companies and representatives throughout the world

A catalogue record for this book is available from the British Library.

ISBN 0-333-68248-3

First published in the United States of America 1998 by
ST. MARTIN'S PRESS, INC.,
Scholarly and Reference Division,
175 Fifth Avenue, New York, N.Y. 10010

ISBN 0-312-21038-8

Library of Congress Cataloging-in-Publication Data
Bashford, Alison, 1963–
Purity and pollution : gender, embodiment, and Victorian medicine
/ Alison Bashford.
p. cm.
Includes bibliographical references and index.
ISBN 0-312-21038-8 (cloth)
1. Social medicine—History—19th century. 2. Body, Human—Social
aspects—History—19th century. 3. Women in medicine—History—19th
century. 4. Sex role—History—19th century. 5. Gender identity-
-History—19th century. 6. Feminist theory. I. Title.
RA418.B367 1998
610'.9'034—dc21   97-28323
                              CIP

© Alison Bashford 1998

All rights reserved. No reproduction, copy or transmission of this publication may be made without written permission.

No paragraph of this publication may be reproduced, copied or transmitted save with written permission or in accordance with the provisions of the Copyright, Designs and Patents Act 1988, or under the terms of any licence permitting limited copying issued by the Copyright Licensing Agency, 90 Tottenham Court Road, London W1P 9HE.

Any person who does any unauthorised act in relation to this publication may be liable to criminal prosecution and civil claims for damages.

The author has asserted her right to be identified as the author of this work in accordance with the Copyright, Designs and Patents Act 1988.

This book is printed on paper suitable for recycling and made from fully managed and sustained forest sources.

10 9 8 7 6 5 4 3 2 1
07 06 05 04 03 02 01 00 99 98

Printed and bound in Great Britain by
Antony Rowe Ltd, Chippenham, Wiltshire

For Tessa and Oscar

Contents

List of Abbreviations viii

List of Illustrations ix

Acknowledgements x

Introduction xi

1 Sanitising Spaces: *The Body and the Domestic in Public Health* 1

2 Female Bodies at Work: *Narratives of the 'Old' Nurse and the 'New' Nurse* 21

3 'Disciplines of the Flesh': *Sexuality, Religion and the Modern Nurse* 41

4 Pathologising the Practitioner: *Puerperal Fever in the 1860s* 63

5 Feminising Medicine: *The Gendered Politics of Health* 85

6 Dissecting the Feminine: *Women Doctors and Dead Bodies in the Late Nineteenth Century* 107

7 Sterile Bodies: *Germs and the Gendered Practitioner* 127

Notes and References 149

Bibliography 173

Index 183

List of Abbreviations

BL	British Library
BMJ	*British Medical Journal*
GLRO	Greater London Record Office
NAPSS	National Association for the Promotion of Social Science
TNAPSS	*Transactions of the National Association for the Promotion of Social Science*

List of Illustrations

1.1	The Ladies' Sanitary Association 1866 (The Wellcome Institute for the History of Medicine Photographic Collection)	13
2.1	'Fifty Years' (*The Nursing Record and Hospital World*, 20 December 1888)	29
6.1	Dissection of the Male Perineum (D. J. Cunningham, *Manual of Practical Anatomy*, Young J. Pentland, Edinburgh and London, 1893, p. 348)	118
7.1	'Sterilization of the Hands and Skin' (*British Medical Journal*, 30 September 1905, p. 783)	143

Acknowledgements

My thanks to Barbara Caine who first interested me in the history of feminism and who continues to offer guidance, criticism and suggestions at a moment's notice. At the University of Sydney, Glenda Sluga is a most valued friend and reader, and I hope she and I stay only a corridor's length apart for many years to come. My immediate colleagues in the Department of Women's Studies, Elspeth Probyn and Gail Mason, create a pleasant workplace. Stephen Garton, Caroline Jordan, Rosemary Pringle and Martha Vicinus offered early suggestions and I thank them for these. Claire Hooker's research assistance was impeccable. Catriona Elder's companionship, humour and intellectual sharpness have sustained me for many years now. Thank you. Emma Partridge has also been an important friend. Nicholas Brunton's extraordinarily good nature, his capacity to both calm and motivate, and his ready willingness to take on the care of the twins in their first year *and* write his own PhD, leaves me constantly impressed and deeply grateful.

I am grateful to the Australian Vice-Chancellors' Committee, whose Bicentennial Scholarship funded research in London. This was extended by a University of Sydney Research Grant. In London, the Wellcome Institute for the History of Medicine was forthcoming with extremely generous resources and office space, and I am grateful for the interest and hospitality with which scholars in the history of medicine are always received.

ALISON BASHFORD

Introduction

'Dirt in its largest sense [is] matter in the wrong place.'
Elizabeth Blackwell, *Scientific Method in Biology*, 1898.

'Dirt is essentially matter out of place.'
Mary Douglas, *Purity and Danger*, 1966.

In an influential report written in 1863, the Medical Officer of the British Privy Council, John Simon, likened hospitals to homes. What he says in this passage, and the language in which he says it, invoke many of the concerns of this book.

> That which makes the healthiest house likewise makes the healthiest hospital, – the same fastidious and universal cleanliness, the same never-ceasing vigilance against the thousand forms in which dirt may disguise itself, in air and soil and water, in walls and floors and ceilings, in dress and bedding and furniture, in pots and pans and pails, in sinks and drains and dust-bins...for the establishment which has to be kept in such exquisite perfection of cleanliness is an establishment which never rests from fouling itself.[1]

Simon was a sanitary reformer. As such, he was obsessed with dirt and cleanliness, with purity and pollution. As this passage suggests, he wanted to bring the idea of domestic cleanliness into hospitals and, conversely, to render the home a place where public hygiene and health were to be systematically pursued. This focus on domesticity meant that women and a particular type of middle-class femininity were centrally implicated in the movement for sanitary reform and public health in Britain.[2] Analysis of this involvement is one focus of this book. But women were implicated in the discursive domain of hygiene and sanitary reform at several other levels. For those familiar with nineteenth-century conceptions of 'woman', John Simon's words are immediately suggestive of the connections between what I call 'sanitarian discourse' and the discourse which constructed the female body. Sanitarian discourse functioned through dichotomous concepts of 'purity' and 'pollution'; what the medical historian Charles Rosenberg has labelled 'morally resonant polarities'.[3] The whole project of sanitary reform invoked a state of cleanliness always defined through the filthiness and impurity it displaced; it was driven by a pollution which continually threatened to recur. It is also the case that in Victorian culture the dichotomy of purity and pollution cohered around, and produced the meaning of 'woman' with particular intensity. Women as a group, as well as individual women, were constructed by middle-class Victorian culture to embody both purity and pollution, to be potentially both madonna and whore, angel and temptress. And like the idea and process of sanitary reform in which the moral and the physical were always conflated, women's purity and impurity were expressed at once morally and physically. While Simon wrote of the house or the hospital as an establishment 'which never rests from fouling itself', he might also have been writing of a dominant nineteenth-century conception of the female body. This body, like domestic and urban spaces, required sanitising.

Historians dealing with nineteenth-century medicine have long written about embodiment, because medicine is inescapably about the human body. Conventional medical historians, feminist historians, postmodern historians have documented the various ways in which the body in health and ill-health has been thought about, even as they have done so from vastly different epistemological positions. By and large however, it has been the patient or the object of medical knowledge and practices which has been thought of as embodied. In this book I examine practitioners of medicine, as well as patients, as embodied subjects. What were the cultural meanings embodied and performed by different practitioners in the Victorian period? How did doctors think about themselves: as fleshly and tactile, sick or well, as carrying a contagion, or as somehow being immune to contagion? How did nurses conceptualise and enact that cleanliness so insistently impressed upon their very bodies and morals as well as their wards and homes? How were different types of practitioners implicated in a discourse and a material practice which was inevitably about the pure and the polluted? In what ways and under what circumstances did the medical gaze turn inwards?

In that both patients and practitioners were embodied, they were also sexed. Such a statement seems all too obvious. Yet in all the recent scholarship on the body emerging from many disciplines and interdisciplinary fields, it is too often implied that human bodies somehow come in one generic type. There has been a rightful insistence on the part of feminist theorists and philosophers that sexual difference and sexual specificity always be recognised and analysed. Bodies – human subjects – do not come in one type. They are always sexed.[4] The different and relational cultural meanings of the femininity or masculinity embodied by a female or male doctor, a female nurse, a midwife or an accoucheur in the Victorian period are a further focus of this book. It has been my aim to write a medical history which is about many types of practitioners, none of whom functioned discursively independently of one another. In particular, it has long seemed to me imperative that male doctors and female nurses be put together, historically speaking. Each was culturally produced with implicit and explicit reference to the other: doctors' masculinity (read dominance, objectivity, scientific authority) was defined in relation to and as complementary to, nurses' femininity (read obedience, emotiveness, religiosity). As one doctor wrote in 1887, 'the nurse was not only the doctor's right hand – she was his *alter ego* – his other self'.[5] Of course, this is not to argue that doctors and especially nurses literally upheld or enacted these characteristics, but that these were dominant ways in which the idea of 'doctor' and 'nurse' were constructed, in turn affecting individual subjectivities and relations. I also devote considerable attention to the analysis of women doctors, midwives and female nurses in relation to each other. The fluidity and complexity of women's place in the scheme of nineteenth-century medicine is often quite misrepresented in the common scholarly insistence on separating historical treatment of nurses, midwives and women doctors.[6] I argue that there were many connections between them. In particular, women doctors and the 'new' nurses who emerged from the 1860s onwards were represented within a similar discourse of women's place in sanitary reform; a discourse shaped by notions of middle- and upper-class women's duties, responsibilities and capacities for philanthropic mission, and their supposed morality, spirituality and religiosity. The figure of the 'female medical practitioner' in the 1860s was ambiguously nurse, midwife and doctor.

This book has emerged from an interest in several fields: the history of Western feminism and of British feminism in particular; nineteenth-century medical history; feminist and cultural theories about the body. It can be located within a longstanding tradition of feminist medical history, much of which has rendered our understanding of medicine in the Victorian period more sophisticated by gendering the field; by analysing, for example, the gendered dimensions of clinical encounters, the vast medico-scientific literature and interest in the nature of 'woman', the investments which a masculine professional world had in retaining control over certain knowledges. Many of the earliest of these histories reflected unashamedly the political concerns of their authors in the 1970s, a moment when the projects of 'women's history' and 'women's health' were being forged and dovetailed. Indeed the investigation of medicine made up a large part of the early field of women's history.[7] These histories were concerned overwhelmingly with woman-as-patient-as-object of a masculine medicine seen to be relentlessly oppressive.

A subsequent wave of feminist medical histories have been inflected by developments in literary theory, critical theory and cultural studies, and have worked from a different understanding of what the historical project is.[8] Louise Newman has written that for such latter historians 'knowledge of the past, of the world, of ourselves, and of sexual difference comes not from reconstructing 'objective' experiences but through analyzing the systems of meaning that make possible and construct those experiences in the first place'.[9] Further, along with important works by feminist philosophers and epistemologists the problematic of 'woman-as-patient' or 'woman-as-object' has been theorised, not simply described as an experience. Women and men, as objects and subjects, did not fit into the objectifying discourse of science and medicine in some sort of arbitrary way, but in a way consistent with, and structured by, longstanding and deeply embedded gendered dichotomies. In countless examples – philosophical, theoretical, historical, material – a dominant (but never uncontested) way of thinking, acting and representing, constructed 'woman' as 'nature' and as 'body' against 'man' as 'culture', 'science' and 'mind'. If woman or even the feminine was the quintessential object of medico-scientific discourse, then man was the subject.[10] Literally, of course, this was not always the case. Indeed, occasionally the position of men and women was reversed. However, as I argue in my analysis of female medical students and their dissection of the male body in Chapter 6, attempts by women to become the active subjects of medicine were shaped and heeded by this gendered scheme. It was all but impossible to contemplate a female practitioner caring for a male patient, not only because of the more obvious issues of Victorian etiquette and sensibilities about interactions between men and women, but also because the medical encounter between patient and practitioner was so firmly one between an implicitly masculine subject and feminine object. It must be said also, that such insights into the gendering of the subjects and objects of science and medicine have a longer history within feminist thought than is usually acknowledged. In Chapter 5 I suggest that a version of this theory was articulated by nineteenth-century feminists, who saw women as victims and objects of an oppressive medicine and who firmly aligned women and female practitioners with the idea of a benign and caring 'nature'. Most of these early feminists wanted to keep the dichotomy which aligned women with nature firmly intact, and to use it to their better their social position. By contrast, the aim of many (but not all) late twentieth-century feminists is to reveal how such a dichotomy functions in order to radically disassemble it and to suggest that it is complicated by all kinds of other subject positions – in particular those of race, class and sexuality.

Much of this book continues a focus on the female body which has characterised medical knowledge itself, as well as feminist critical reflections on this medical knowledge. In Chapter 2 I examine the cultural meanings which female nurses embodied and represented within hospitals and homes. I explore the significance of the idea and reality of the 'old' nurse, seen to be ignorant, filthy, and disordered, and the 'new' nurse, seen to be chaste, pure, and ordered. The story of the shift between types of nurses – told not infrequently as the shift between Dickens' Sairey Gamp and Florence Nightingale – was one which gathered its meaning and significance within ideas about sanitary reform. I argue that the very body of the new nurse, whose 'purity' was at once moral and physical, mapped with considerable resonance onto nineteenth-century ideas about disease and sanitary reform. The new nurse was a sanitised version of the old nurse, a middle-class figure of efficiency, neatness and whiteness. In that ordering, cleansing, purifying and moralising the domestic had come to be so firmly the cultural territory of middle-class women, the new 'lady nurses' and 'lady superintendents' were crucial in the process by which hospitals were modernised and sanitised.

It is no coincidence that the concepts of purity and pollution resonate equally strongly within the domains of Christianity and medicine, of the spirit and the body. In Chapter 3 I look at the ambiguous discursive location of nineteenth-century nursing between religion on the one hand, and rational science and secular medicine on the other. How were these women and the cultural work they performed placed in the large context of the rationalisation and modernisation of institutions and the secularisation of care of the body? In particular, I ponder the significance of female sexuality in this scheme, and argue that the extraordinary discipline of nurses and control of their sexuality, was a major expression of the rationalisation of a troubling pre-modern disorder. Michel Foucault's *Discipline and Punish* offers useful insights into the modes by which this bodily discipline and control operated in the modernisation of nurses and of hospitals.[11] I am also interested in the way in which this process of modernisation, secularisation and rationalisation was not fully realised, indeed was riddled with contradictions. While nurses were major objects as well as agents of modern ordering and discipline, they also continued to be defined and define themselves within a religious discourse. I ask why these quasi-religious figures (who were significantly an innovation in the British Protestant context) entered the domain of medicine precisely when it was undergoing a process of secularisation, that is precisely when medicine was becoming scientific and rational.

Historically and theoretically, masculinity and the male body has been far less scrutinised than femininity and the female body. In the field of medical history, the masculinity of doctors has largely been unstated and unanalysed. Both nineteenth-century male doctors and many twentieth-century historians were/are positioned within the same dominant discourse of gender in which masculinity is 'unseen', 'unmarked', invisible or apparently neutral. A masculine myth of disembodiment has largely shaped both nineteenth-century male doctors' understanding of themselves, as well as historians' writing about them. While whole volumes are waiting to be written on medicine, masculinity and embodiment, I focus specifically on male doctors in two chapters, on two occasions when they were forced to recognise and deal with their own bodies. In Chapter 4 I look at medical discussion of puerperal fever around the 1860s when large numbers of doctors seriously began to consider themselves the carriers of the contagion. This was also a decade which saw an ongoing campaign by feminists and others against accoucheurs as being both morally and physically contaminating; as being morally reprehensible and illegitimately sexual, as well as being the purveyors of disease. In large part, this debate took place with reference to female practitioners – both midwives and the few women who sought conventional medical education. Male doctors, I argue, were 'sexed' in this process. They were spoken of specifically as being men and their masculinity became visible – the object of a pathological scrutiny.

In Chapter 7, I turn to the period between the mid-1890s and the beginning of the twentieth century, when the knowledge and practice of asepsis began to force surgeons again to admit their own status as potential polluters, and to perform new rituals of cleanliness through which their dirty bodies were sterilised. I also argue in this chapter that the supposed 'revolution' which asepsis brought to medicine and health-care, was certainly revolutionary for most male doctors, but was not for female nurses, whose work, sense of embodiment and cultural representation had been forged within a language and practice of absolute purity for decades.

As a field of sometimes quite powerful knowledge-production, medical history generally has a massive investment in the perpetuation of problematical modern visions of scientific progress, partly because this history-writing is often institutionally bound up with medical science itself, rather than being located within the humanities or social sciences, or even as an interdisciplinary project. It should already be more than apparent that I question any idea that the history of medicine has been a history of progressively better practices and more true knowledges. Rather, I write from an epistemological position which refuses fixed meanings and essential truths; a position which suggests that scientific knowledge actively constructs and produces its object of study, rather than simply reveals it. The Victorian era saw many different conceptions of the body in health and illness come, go and coexist. It is important to work within a theoretical framework which rejects the idea that *one* of these conceptions was 'right', the suggestion that one of them (in most accounts of course, the conception of illness produced through germ theories) held and finally indicated accurately the essential truth of the sick body. Medical knowledge is always socially constructed and always culture-bound: it is never simply 'true'. In this sense, *Purity and Pollution* is a cultural history.

1 Sanitising Spaces: The Body and the Domestic in Public Health

The mid-nineteenth-century movement for sanitary reform has long been of interest to medical historians. The ways in which this movement was shaped by a gendered politics of health, however, has been little discussed. Given the sheer frequency with which women's place in sanitary reform was articulated in the widest possible range of nineteenth-century texts and discursive practices, this lack of interest is curious indeed. Dr Benjamin Ward Richardson, for example, wrote in his 1880 lecture 'Woman as a Sanitary Reformer' that '[i]t is in those million centres we call the home that sanitary science must have its true birth'.[1] Needless to say, it is precisely this connection between domesticity and femininity which has placed the topic 'woman as sanitary reformer' outside the interests of a largely masculinist tradition of public health histories. However, as I argue in this chapter, and as was more than apparent to nineteenth-century sanitary reformers themselves, 'public' health was secured largely in 'private', that is in domestic spaces. It seems to me that in a range of ways, sanitary reform was never really settled as a masculine or feminine domain.

From the earliest formulations of the meaning of public health in the 1830s, through the 1860s and 1870s when Richardson was influential, and arguably into the early twentieth century, several issues meant that women, in a variety of capacities, were involved in sanitary reform. First and most important was this centrality of the domestic; in particular the domestic space and culture of working-class homes. Second, sanitary reform always involved some sort of moral reform, which rested upon theories of disease which conflated physical and moral cleanliness and health, and perhaps more pertinently, physical and moral dirtiness and ill-health. Given the particular responsibility for morality and moral education which Victorian culture ascribed to middle- and upper-class women, their involvement should come as no surprise. The workings of this morality also came into play in the class dimensions of the project of sanitary reform. 'Sanitising' domestic space also meant 'moralising' it into middle-class values. Through the imposition and in part the acceptance of fundamentally middle-class codes of morality, working-class suburbs, homes, families, lives, and bodies were to be ordered and sanitised in both physical and cultural senses. The significance of both the domestic and the moral dimensions of sanitary reform meant that middle- and upper-class women located themselves in the discourse with considerable authority, an authority also dependent on their position within philanthropic culture.

Historical analysis of sanitary reform is often framed within debate about contesting theories of disease; the supposed paradigm shift from miasmatic theories to germ theories which occurred from the mid to the late nineteenth century.[2] Insofar as gendered readings of sanitary reform have been forthcoming, they also have been made to fit this disease-theory model, and with some difficulty. Examining English ladies' sanitary associations, Perry Williams has argued that '[w]omen's work in sanitary reform... did not continue much beyond the mid-Victorian period' and that 'with the increasing adoption of germ theory, the importance and even the possibility of women's sanitary work was severely diminished'.[3] While Williams's article points rightly to the importance of women's place in sanitary reform, the assertion that this came to an end in the mid-Victorian period is unfounded. Nancy Tomes, to the contrary, has argued that 'domestic hygiene' was influential well into the twentieth century, albeit in the American context.[4] Tomes's article points convincingly to the connection between sanitary reform and what has become known as the 'new public health' based on germ theories, although the gender and class politics and dynamics of these processes, if implicit in her work, would appear to be more or less peripheral. In this chapter I extend some of these insights on gender and sanitary reform. First, I sketch out the broad context of nineteenth-century public health including an outline of dominant theories of disease which will be referred to throughout the book. I will describe some of the class and gender dynamics driving the project of public health, which made it possible for domesticity and femininity to be so centrally invoked in the cause of sanitary reform and which gave rise to women's sanitary associations. And in the final section I begin to explore the significance of the body. In discourse on public health, there was a distinct conflation of ideas around the body and the domestic. Sanitary reformers wrote of bodies as houses to be kept clean and pure, and of houses and bodies to be purified and sanitised along the same hygienic principles.

THEORIES OF DISEASE AND PUBLIC HEALTH

The movement for sanitary reform in England emerged in the 1830s and 1840s. In part, it was a response to public health problems, such as water supply and sewage, which arose with industrialisation, urbanisation and population growth, as well as epidemics of infectious diseases which had not been encountered before in scale, and in some cases, notably cholera, in type.[5] The whole question of overcrowding – the increasing number of bodies in restricted urban spaces – became quite central. Campaigns for public health involved controlling the material and cultural conditions of the urban working class through established philanthropic structures and increasingly through state institutions. Several major public health acts were passed in Britain through the 1840s and were expanded and vigorously implemented in the following decades. The domain of public health and sanitary reform was a major manifestation of, and site for bureaucratisation and the growth of state administration.

Sanitary reform was never an apolitical process, simply seeking humanely to create the conditions for greater health. It was, for example, fundamentally bound up with the emergence of the problem of pauperism, as it was constructed by a new social science, and with class issues surrounding the administration of the New Poor Law.[6] Additionally, sanitary improvement and the new discourses which quantified, recorded and categorised disease were as much about protecting the health of the middle and upper class, as about undertaking benevolent work.[7] Mary Poovey has also argued that sanitary reform, or more particularly Edwin Chadwick's 1842 *Report on the Sanitary Condition of the Labouring Population of Great Britain*, involved a process of domesticating (male) working-class culture to middle-class standards, in which men were to locate themselves primarily in the home, rather than in a range of public places where same-sex political alliances were built. For Poovey, the significant implication of this was that it limited 'the ability of working-class men to organize themselves into collective political or economic associations'. In this plan of 'improvement', implemented by bureaucrats such as Chadwick, working-class women were 'crucial to the domestication, individualization, and (by extension) depoliticization of working-class men'.[8] A new politics of health was at work, organised and implemented through the interventions of a range of emerging local health practitioners – doctors, sanitary inspectors, factory inspectors and so on. The politics of this sanitary reform, also called a 'sanitary science', were succinctly articulated by Edwin Chadwick in his 1860 'Address on Public Health'.

> To those specially interested in reformatories and the repression of crime, to those specially interested in the repression of pauperism and mendicity, and the relief of sickness and destitution, as well as those generally interested in the repression of social disorder, I submit, that the primary preventive measure, for which we ought now to labour in common, is to ensure to the rising generations a sound mind, having for its necessary foundation a sound body... under the direction of sanitary science.[9]

Michel Foucault described such a politics as involving the 'consideration of disease as a political and economic problem for social collectivities which they must seek to resolve as a matter of overall policy'.[10] In 'The Politics of Health in the Eighteenth Century', Foucault wrote on the importance of the health of a population for a new type of political power. This was not simply

> a matter of offering support to a particularly fragile, troubled and troublesome margin of the population, but of how to raise the level of health of the social body as a whole. Different power apparatuses are called upon to take charge of 'bodies', not simply so as to exact blood service from them or levy dues, but to help and, if necessary, constrain them to ensure their own good health. The imperative of health: at once the duty of each other and the objective of all.

He wrote that a '"medico-administrative" knowledge begins to develop concerning society, its health and sickness, its conditions of life, housing, and habits, which serves as the basic core for the "social economy" and sociology of the nineteenth century'.[11]

Those who sought sanitary improvement through new principles of engineering, building and architecture, through urban planning and broad-sweeping public health legislation, and those who wrote texts and tracts, lectures and pamphlets on public health were often quite explicit about the connection between moral and physical wellbeing. Chadwick's *Report on the Sanitary Condition of the Labouring Population of Great Britain* pointed to the 'coincidence of pestilence and moral disorder' and largely ascribed 'immorality' to the want of adequate sanitation.[12] In his book on Victorian public health, *Endangered Lives*, Anthony Wohl has written:

> [P]ublic health, like so many other social reforms and endeavours, took on the form of a moral crusade... the most widely held of Victorian social doctrines was that physical well-being and a pure environment were the essential foundations for all other areas of social progress. In short there could be no moral, religious, or intellectual improvement without physical improvement.[13]

Wohl is right to attribute such significance to the close relations between physical and moral issues. However, his description fails to do justice to the way in which, in the sanitarian mind, physical and moral issues were not simply related as cause and effect, one as the foundation for the other, so much as absolutely intertwined. As the National Association for the Promotion of Social Science was told in 1858, 'the physical and the moral... act and re-act upon one another in a way quite indivisible'.[14]

This conflation of morality and physicality was underpinned by miasmatic theories of health and ill-health. From the Greek word 'to pollute', miasma was both a theory of disease causation and a description of 'filth' or polluting matter. Disease was understood to be a response to decomposing, putrefying matter in the surrounding environment – human waste, accumulation of dirt, stagnant water, foul air. The latter was understood as the main medium of transmission, although the distinction between transmission and causation was not clear, nor indeed was it seen to be particularly significant.[15] The management of airflow through ventilation was accorded extraordinary significance in the planning and management of private homes as well as institutions such as gaols and especially hospitals. The opening and closing of windows and the use of fires were seen to 'draw off' miasmas or to prevent their entry from outside. Florence Nightingale commanded windows to be opened: '[F]irst rule of nursing, to keep the air within as pure as the air without.'[16] For others in these years, however, it was common practice to keep all windows shut so as to avoid the entry of miasma from external sources.[17] Alternatively, huge fires would be maintained in the sick-rooms of private houses or in wards to draw the miasma through the chimney.[18] Developing scientific formulae about airflow and cubic space of air per person was a major concern of sanitarians, designers and physicians in the mid-nineteenth century. Implementing such theories was a major function of nurses in hospitals and women in homes. It was a serious business indeed. One authority stated categorically that '[c]ontamination and deterioration of the air of hospitals produces erysipelas',[19] while another specifically placed the responsibility for mortality from puerperal fever in hospitals *'according to the efficiency or inefficiency of the ventilation'*.[20] Florence Nightingale's 'pavilion' plan hospitals were designed to have each ward surrounded by clean air, and a modified version of her plan became standard for hospital design in the late nineteenth century.[21]

Edwin Chadwick's famous statement that 'all smell is disease' sums up the miasmatic view of disease.[22] Clearly, such material and empirical knowledge did not privilege those with medical expertise. Doctors were not the automatic authorities. Public health, sanitary reform and miasmatic theories of disease allowed for other sorts of expertise based on practicality and common sense, as well as on moral and domestic authority. For example, in an 1871 letter, Nightingale wrote of cases of pyaemia, or blood poisoning, at St Thomas' Hospital and was able to state, quite categorically: 'Mrs. Wardroper & I have as I believe discovered the cause.' Nightingale's description is a clear statement of the direct material link in the sanitarian model between filth, smell and disease:

> *All* the refuse of the Hospital... had been ever since the occupation conveyed to a dust-hole between No. 7 Block & the *Steward's* house – & only emptied twice a week.
> The 8 (*first*) fatal cases of Pyaemia occurred in *No.7 Block* – & in no other.
> A few weeks ago the *Steward* (not liking the smell) transferred the dust-hole to the *basement under* No.3 Block – & close under the windows of our Probationers' Home & of Mrs. Wardroper.
> A fatal case of Pyaemia appeared immediately in *No.3 Block* – & several cases of illness among the Prob'rs... *No* case of Pyaemia has appeared in No.7 Block since the dust-hole was removed from there.[23]

Such categorical claims about disease aetiology and prevention were entirely possible for non-doctors within sanitarian discourse. Indeed the initial impetus for sanitary reform in England came from outside the medical profession altogether, from utilitarians and political economists, and was carried through by non-medical bureaucrats such as

Chadwick.[24] Yet by the 1860s and 1870s there was a contention for authority over public health between sections of the medical profession on the one hand and non-medical sanitarians, such as Chadwick and Nightingale, on the other. A new generation of doctors was emerging, broadly sympathetic to the latest versions of contagionist theories in which disease was thought to be caused by specific particles.[25] Charles Rosenberg's description of such contagionist theories as 'morally random' precisely encapsulates their difficulty for sanitarians such as Nightingale who attempted to create 'moral universes' in order to combat disease. Rosenberg argues that Nightingale and her ilk viewed the world 'in morally resonant polarities: filth as opposed to purity, order versus disorder, health in contradistinction to disease'.[26] Increasingly under challenge from men educated in contagionist theories and later germ theories, it was to become less possible for non-medical women like Nightingale to claim to have 'discovered the cause' of any disease. In the middle of the century, though, there was as yet no particular authority attributed to contagionist theories as opposed to sanitarian theories. Certainly there was no self-evident truth in the former, scientific or otherwise, which guaranteed its success. If anything, the sanitarian, anti-contagionist, anti-reductionist position, typified by Nightingale, was understood to threaten the monopoly on issues of health sought by a new generation of doctors, who located the cause of disease in microscopic organisms.

The extent to which sanitarian ideas undercut the social position of such doctors is evident in Dr Benjamin Ward Richardson's 1875 publication *Hygeia: A City of Health*. This immensely popular publication, dedicated to Edwin Chadwick, was first read as a paper to the health department of the Social Science Congress. It described in minute detail Richardson's picture of a utopian city built from the latest sanitarian principles: small populations; no tall buildings which would restrict sunlight; no underground rooms of any sort; no carpets to harbour dirt; no public laundries. Rather, *Hygeia* would boast 'swimming baths, Turkish baths, playgrounds, gymnasia, libraries, board schools, fine art schools, lecture halls' as well as a 'large ozone generator'.[27] As will be seen, *Hygeia* was a text which also demonstrated the particular space within the discourse of sanitary reform which women could claim and manipulate. Although an eminent physician himself, the message Richardson put forward in *Hygeia* was that doctors would have less, not more control and status in such a sanitarian city. *Hygeia* was caricatured in *Punch* as putting doctors out of work entirely.[28] Richardson was heavily critical of the specialising trend of medical treatment and described his 'model hospital for the sick' thus:

> They are small, and are readily removable. The old idea of warehousing diseases on the largest possible scale, and of making it the boast of an institution that it contains so many hundred beds, is abandoned here... The still more absurd idea of building hospitals for the treatment of special organs of the body, as if the different organs could walk out of the body and present themselves for treatment is also abandoned... The officers are called simply medical officers, the distinction, now altogether obsolete, between physicians and surgeons being discarded.[29]

As Richardson suggested in *Hygeia*, there was a trend for sections of the medical profession to think of the body as reducible to its different parts. In Nightingale's and Richardson's sanitarian minds, the body should not be thought of as separate organs, but as a whole, interactive organism, which required constant internal and external balancing and regulating. This is well illustrated by the title of an 1858 text, 'Hygiene or Health as Depending Upon the Conditions of the Atmosphere, Foods and Drinks, Motion and Rest, Sleep and Wakefulness, Secretions, Excretions and Retentions, Mental Emotions, Clothing, Bathing &c.'[30] In this dominant but contested understanding, the healthy or sick person was not medically or culturally conceptualised as separate from what surrounded them, but rather as part of an emotional, environmental, even architectural context which could be either therapeutic or productive of illness.

DOMESTICITY IN THE DISCOURSE: PUBLIC HEALTH IN PRIVATE

It is now something of a commonplace in modern Western feminist history to point out that as men gained more political and economic power in industrialised societies, middle-class women were culturally compensated, so to speak, with a heightened moral and spiritual place. Women's special moral sensibility and religiosity were at the core of middle-class femininity as it was constructed and contested through the nineteenth century. In England, an evangelical revival, the separation of home and work, and the creation of gendered (but never really separate) 'public' and 'private' spheres were connected developments. Women's spirituality and supposedly superior morality and religiosity came to be centred on ideas about the domestic sphere: the sanctity of the home; the moral authority of women within it; the different and special place this gave women in the overall social structure.[31] Mary Poovey has summarised these developments thus:

> Instead of being articulated upon inherited class position in the form of noblesse oblige, virtue was increasingly articulated upon gender in the late eighteenth and early nineteenth centuries. As the liberal discourse of rights and contracts began to dominate representations of social, economic, and political relations, in other words, virtue was depoliticized, moralized, and associated with the domestic sphere, which was being abstracted at the same time – both rhetorically and, to a certain extent, materially – from the so-called public sphere of competition, self-interest, and economic aggression. As superintendents of the domestic sphere (middle-class) women were represented as protecting and, increasingly, incarnating virtue.[32]

This constellation of ideas – domesticity, morality, religiosity, bodily cleanliness, purity and virtue – shaped the possibilities for women's involvement in sanitary reform and public health. The new politics of health was not only being negotiated in local relations between men, as implied by Foucault, but also across the sexes, and between differently placed women.

From the outset, domestic space, domestic issues, domestic relations figured prominently in sanitary reform and public health. Chadwick's *Report on the Sanitary Condition* hints at what was to become a central strand within the subsequent sanitary reform movement: the reconstruction of working-class domestic space as feminine within the ideology of 'separate spheres' and the influence of middle-class women on working-class women to sanitary and public health ends. Chadwick wrote: 'One of the circumstances most favourable to the improvement of the condition of an artisan or an agricultural labourer, is his obtaining as a wife a female who has had a good industrial training in the well regulated household of persons of a higher condition.'[33] Domestic training and the influence of middle-class family life, especially that of middle-class women, were seen to be curative of social, physical and moral ills: 'The females are from necessity bred up from their youth in the workshops... The minds and morals of the girls become debased, and they marry totally ignorant of all those habits of domestic economy which tend to render a husband's home comfortable and happy; and this is very often the cause of the man being driven to the alehouse.'[34] Part of the nineteenth-century process of cleansing and of sanitising, of the 'purification and improvement' of working-class districts,[35] was to be the extension of the idea of the middle-class family, and of the family home kept physically and morally clean by the wife and mother. While by the early twentieth-century sanitary knowledge became increasingly the domain of professional experts both male and female, there was an earlier assumption that at least some aspects of such knowledge were harboured within middle-class culture in general, and especially in middle-class feminine domestic culture.

From the 1860s, the home and its management were increasingly constructed in sanitarian terms and along scientific principles. Intimately related was a process whereby hospitals began to be reconceptualised in domestic terms. As homes came to be thought of scientifically, as places where the public health was to be systematically pursued, so hospitals came to be thought of domestically. Central to this was the new idea of trained female nurses, governed by middle-class lady superintendents. The boundaries between institutional and domestic sites for the production of public health became decidedly fluid. Sanitary principles of design, of management and of practice applied to hospitals and to homes quite similarly: they became mutually referential. Chadwick wrote that 'the sanitary measures necessary to ensure successful treatment in hospitals, may be stated in respect to common dwellings'.[36] All the elaborate theories about air circulation and the desired cubic space around each body applied as much to homes as they did to hospitals. In the same way that there was a rush of publications about hospital construction from the 1860s and 1870s, so there was publication after publication about the sanitary construction, ventilation and cleaning of homes.[37] Sickrooms in institutions and in private homes became subject to the same principles of arrangement, order, cleanliness. And so, for example, Nightingale's *Notes on Nursing*, intended for women in a domestic and familial setting, could slip with ease into its role as the most influential nineteenth-century text for women working in hospitals.[38]

Sanitary reform, then, increasingly resonated with the cultural meanings of domesticity. By invoking homes, it invoked women's place within homes: an upper middle-class household headed by a lady with her hierarchy of domestics to maintain standards of health, order and cleanliness; a working-class home aired and scrubbed by a fastidious housewife. Richardson's writings strongly asserted women's place within the world of sanitary reform, not least as housewives: 'Long before the word Sanitation was heard of, or any other word that conveyed the idea of a science of health, the good, cleanly, thrifty housewife was a practical sanitary reformer.'[39] The ideal housewife, according to Richardson, was one who 'would make the very act of cleaning and cleansing clean – the scullery, the landing, the bathroom, the laundry – the cynosures of the household'.[40] For decades, British commentators referred to Richardson's views on the importance of women in sanitary reform, constantly reinscribing the feminine within 'hygiene'. As late as 1911 a textbook on hygiene began with a reference to Hygeia. Hygiene, the author wrote, was 'a very ancient science. It derives its name from Hygieia (*sic*), daughter of Aesculapius, who was himself a celebrated physician. She was greatly honoured by the Ancients as Goddess of health'.[41] The construction of the pursuit of hygiene, health or preventive (as opposed to curative) medicine as a feminine enterprise, was common throughout the period under review.

Needless to say, the centrality of domesticity was both enabling and constraining for women, operating differently for women from different classes. While there were always restrictive tendencies in these areas of women's activity, given the firm connection between womanhood and the home which sanitary reform sustained, there were also significant possibilities for the expression of a certain authority on health issues and the development of particular activities outside the home and in spheres of paid employment. In England in the 1860s, activism around sanitary reform was closely connected with an emerging organised feminism. The feminist *English Woman's Journal* was described as the organ of the Ladies' Sanitary Association, and certainly many of the most prominent feminist figures, especially Bessie Rayner Parkes, took up sanitary reform as a pressing social issue which demanded women's involvement.[42] On the one hand, sanitary reform created the possibility for a whole range of new paid and unpaid occupations for working-class and middle-class women. On the other, it functioned radically to limit the scope of such occupations.

Women's place within the philanthropic arena of sanitary reform was given formal shape through women's sanitary organisations. Women in England in the late 1850s and 1860s organised energetically and effectively around sanitary issues. As with all separate female organisations, sanitary organisations offered both the possibility of developing a woman-centred and even feminist agenda, as well as providing a structure for the constraint of women within a restricted sphere of commentary, influence and activity acceptable to a patriarchal world. In 1857 the Ladies' National Association for the Diffusion of Sanitary Knowledge was established in London. It emerged from the National Association for the Promotion of Social Science (NAPSS). The Ladies' Sanitary Association, as it came to be called, saw itself as pursuing the most important objects of the male-dominated NAPSS: 'the diffusion of sanitary knowledge, and the promotion of the idea of physical education, especially among the working classes.'[43] These women found authority to speak on sanitary issues because they concerned other women, because they concerned working-class domesticity and because sanitary issues were moral issues. In an 1858 paper on the Ladies' Sanitary Association, Mary Anne Baines wrote that '[a]lthough the primary and more direct object... is to improve the physical condition of the poor, another benefit is likely to result, not less important in its character – an improvement in the moral and social condition of the working-classes, brought about through the natural relation that exists between the physical state and the moral condition, so that a correction of the ills attached to the one bespeaks a reformation of the evils attending the other'.[44] Indeed, women's involvement was sometimes articulated as '*moral* sanitary reform'.[45]

The Ladies' Sanitary Association undertook activities which became the pattern for subsequent associations of women involved in sanitary reform. The publication and distribution of tracts was the single most important function. Some of these tracts aimed to educate women about their own bodies: 'Wonders of Bodily Life', 'Evils Resulting from Rising too Early After Childbirth', 'The Sanitary Duties of Private Individuals'. Others dealt with the application of sanitary principles to the home, focusing primarily on ventilation: 'Mischief of Bad Air' or 'The Worth of Fresh Air', for example. Still others focused on women's newly significant responsibilities as wives and mothers, and as creators of the physical and moral environment of the home. The roots of early twentieth century domestic science and 'scientific motherhood' are readily discernible in tracts from the early 1860s such as 'The Mother', 'Bride's New Home' and 'A Model Wife'.

Dissemination of religious and physical instruction often occurred together – to the extent that the new tracts by the Ladies' Sanitary Association were often bound with religious tracts and distributed

LADIES' SANITARY ASSOCIATION.

RULES OF THE LEEDS BRANCH.

RULE I.—That a Leeds Branch of the LADIES' SANITARY ASSOCIATION (now affiliated with the National Association for the Promotion of Social Science) be formed, for the diffusion of Sanitary knowledge and the promotion of Sanitary measures amongst the homes of the poorer classes, through the co-operation of the Clergy and Ministers of religion, Medical Profession, District Visitors, Bible-women, and all others interested in the welfare of the poor.

II.—That any person paying a Life Subscription of £10, or an Annual Subscription of 10s., be a Member. Also, that any person paying an Annual Subscription of 5s. be a Member; and that the payment of £5 be equal to an Annual Subscription of 5s. That the Subscriptions be due in January each year.

III.—That the Rules of the Ladies' Sanitary Association be adopted by the Branch as far as practicable; and that its business be conducted by a Committee of six or more Ladies, with the Treasurer and Honorary Secretary; three to form a quorum.

IV.—That each Member be supplied with a number of tracts for distribution, not exceeding in value one-third, or at most one-half, of the amount subscribed by such Member.

V.—That grants of Sanitary articles be made by the Committee and Honorary Secretary as far as the funds of the Branch will allow; and that such grants be distributed by a Member of the Branch, or some person deputed by the Committee or Honorary Secretary; and that all recommendations for grants be signed by two Members of the Association.

VI.—That £1 1s. be remitted to the Parent Association for the Annual Subscription.

VII.—That the Accounts be audited yearly, and the Balance Sheet printed with the Annual Report.

FORM OF REQUISITION FOR A GRANT,

REQUIRING TWO SIGNATURES.

We, the undersigned Members of the Association, recommend a grant of Sanitary articles be made to

Signed by

To the Hon. Secretary of the
Ladies' Sanitary Association,
19, Park Place, Leeds.

LADIES' SANITARY ASSOCIATION.

PROSPECTUS OF LECTURES FOR 1866.

The following gentlemen have kindly promised to give Lectures on Sanitary subjects during the current year:—

Dr. GREENHOW, M.D.	"On the Atmosphere."
Dr. ALLBUTT, M.D. (Cantab.)	"On the Causes of Epidemies, and the Best Means of Preventing them."
Mr. IKIN	"On the Means of Preserving Health and Prolonging Life."
Mr. H. MILES ATKINSON	"On How to make our Homes Healthy and Comfortable."
Mr. SCATTERGOOD	"On the Air we Breathe in Large Towns."
Mr. SEATON	"On Personal and Home Cleanliness, with a View to the Prevention of Disease."
Mr. WILLIAM HALL	"On What to Eat, and When to Eat it."
Dr. JAMES BEATTHWAITE	"On Cleanliness."
Mr. JOHN NUNNELLY	"On the Circulation of the Blood, and the Use of the Lungs."
Dr. R. T. LAMB	"On the Air."
Mr. FAIRLEY, F.C.S., Lecturer on Chemistry	"On Water: its Composition, Uses, and Impurities."

NOTE.—Some of these Lectures will be given in the Working Men's Hall; others in different parts of the town. The time and place will be duly announced.

Figure 1.1 Pamphlet from the Ladies' Sanitary Association, Leeds Branch 1866

Source: Courtesy of The Wellcome Institute Library, London

through the networks of bible-women already established.[46] In one address on public health to the NAPSS, it was stated that '[i]n the useful details of domestic hygiene women are the best teachers, and it is gratifying to know that in London female visitors are disseminating such physical instruction at the same time that they are reading the Bible'.[47] Along the lines of upper working-class bible-women, the Ladies' Sanitary Association employed 'Sanitary Home Missionaries' in the belief that little would be achieved without 'oral and practical teaching and personal influence'. Like the bible-women, the female sanitary missionaries were to be 'intelligent women of the working classes' to be trained, and to instruct women in 'domestic sanitary science'. Working-class women were seen to have an empathy and an understanding with the poor which lady visitors could never attain.[48]

One of the earliest objectives of the Association was to bring schoolmistresses and teachers of working-class girls into institutions for free courses on sanitary and health issues, especially those pertaining to care of infants and children, in order that they might incorporate this into their own teaching. The same institutions, it was planned, would run courses for 'private governesses and other ladies'.[49] However, the Ladies' Sanitary Association's main educative role, apart from its tracts, was public lectures, usually given by medical men. Susan Powers, secretary of the Association in 1860, wrote with some pleasure of a group of working women suggesting to one such lecturer the impracticability of his recommendations:

> a lecture was delivered ... by an eminent medical gentleman, to an audience of poor women in one of the worst parts of London. They were very attentive, and made remarks full of practical wisdom, and good sense. They expressed great approbation of the lecturer's instructions; but they clearly showed him that, in more than one particular, his directions had the grave defect of utter impracticability. It was really a most edifying mutual instruction class.[50]

By including this episode in her article, Powers implied that no matter how expert medical men might consider themselves in sanitary matters, women of all classes would always hold an important body of practical domestic knowledge.

As in other areas of social activism in the mid- and late-nineteenth century, middle-class women claimed that it was their right and responsibility to be involved in sanitary reform largely because it involved other women, and issues of domesticity and the family, but also because there was a strong belief that class relations were less rigid between women than between men. As Jane Lewis has commented, 'ladies were better placed to break social ground between the classes'.[51] The Ladies' Health Society of Manchester and Salford, established in 1861, was convinced that ladies were 'comparatively unfettered by the vexed relations between labour and capital' and that with 'their more ready sympathy and common interests with all other women, they would begin hopefully where men would have little chance'.[52] It is important to understand the extent to which women's philanthropic culture was taken as a reforming contribution and intervention, particularly with respect to state-run institutions – the workhouses and infirmaries which were the result of the 1834 Poor Law Amendment Act.[53] Women philanthropists often spoke from a position which opposed what they saw as the dehumanising aspects of institutionalisation and bureaucratisation. In the dominant public discourse, and in terms of the social work which was seen to derive from social science, upper and middle-class women were perceived to humanise and moralise a whole set of relations: relations between classes; between middle-class men and working-class women and children; between religious and scientific social work; between the state and working-class people in workhouses and workhouse infirmaries. Importantly, as Eileen Yeo has pointed out, such women were seen to have the capacity to moralise the new professions and the work of the medical profession in particular.[54] The politics of nineteenth-century philanthropy, charity and social/sanitary science was deeply gendered, as well as deeply classed; the dynamics, interests and manifestations of which were by no means necessarily commensurable. Thus, for example, while Nightingale can be located well within a model of modernisation built around an idea of the social control of the working class and of non-European people, she also subverted many of the gendered controls established by the masculinist society in which she lived.[55] As Mary Poovey has argued, this was done not through overturning feminine discourses, but through proclaiming them, and exploiting the contradictions inherent in the 'domestic ideal' of woman.[56]

In addition to the particular nature of class relations between women, it was commonly agreed that women only could deal adequately with personal issues of hygiene and with the growing concerns around maternal and infant care. A set of arguments developed which detailed the specifics of sanitary work in domestic and personal contexts, which discounted men's and called for women's participation.

In women's writing, the whole domain of sanitary reform was often conceptualised in gendered public and private terms. Outside the house was the responsibility of men; inside the house – significantly a domain which included the bodies of the house-dwellers – was the responsibility of women. Quoting Anna Jameson, Susan Powers wrote in her 'Details of Woman's Work in Sanitary Reform', that such work would combine 'in due proportion the masculine and the feminine element'.[57] Elsewhere, Powers wrote:

> Our legislators may frame wise Acts of Parliament, enforcing sanitary regulations; but some of the greatest causes of disease and death lie quite beyond legislative interference, and can only be removed by woman's agency. The 'Health of Towns Act' may ensure good drainage and water-supply, pure air, and other important external sanitary requisites; but till every woman frames a Health of *Homes* Act, and becomes a domestic 'officer of health', none can ensure that the pure air shall ever be breathed, the good water ever be sufficiently used, or other sanitary conditions ever be fulfilled in-doors.[58]

The idea of women as 'domestic officers of health' was realised to an extent which Susan Powers, writing in the 1860s, could not have imagined. Yet, in the philanthropic arena, and especially as the pursuit of public health became built into state structures, it was to prove extraordinarily difficult to prise apart women's involvement in the domain of health from this grounding as domestic authorities.

BODILY HYGIENE/DOMESTIC HYGIENE

In sanitarian discourse a network of ideas around bodies, homes and hospitals collapsed into one another. Bodies themselves were increasingly conceptualised as domestic spaces and sanitary principles of cleanliness, order and hygiene applied at once to hospitals, homes and bodies. I quote again from John Simon's 1863 report to the Privy Council:

> [F]or the establishment which has to be kept in such exquisite perfection of cleanliness is an establishment which never rests from fouling itself; nor are there any products of its foulness – not even the least odorous of such products, which ought not to

be regarded as poisons. Above all, this applies to the fouling of the air within hospital wards by exhalations from the persons of the sick. In such exhalations are embodied the most terrible powers of disease... and any air in which they are let accumulate soon becomes a very atmosphere of death.'[59]

In such sanitarian and miasmatic conceptualisations of health and disease, bodies and buildings were mutually affective. The construction and cleanliness of houses and hospitals, and the interaction between dwellings and the bodies within them were seen to be crucial and potentially productive of disease. The lack of distinction between physical environment, bodily condition, and moral wellbeing meant that the cleanliness or dirtiness of any of these elements could influence the other. So, a dirty or badly ventilated dwelling, be it a home or a hospital, threatened the onset not only of physical disease, but of declining moral status. Equally, dirt could emanate from an unhygienic body or an immoral body and 'foul' the atmosphere of a room or a building.

Chadwick gave many examples of the way in which occupants could be physically and morally affected by the house or room in which they lived: the way in which people became 'of a piece' with a dwelling. One young woman cited in the *Report on the Sanitary Condition* had been brought up as a female servant in a middle-class household and was taught 'the habits of neatness, order, and cleanliness'. After marrying, she lived in the type of insanitary cottage which Chadwick wanted to condemn: 'But what a change had come over her! Her face was dirty, and her tangled hair hung over her eyes... everything indeed about her seemed wretched and neglected... Her condition had been borne down by the condition of the house.'[60] Another female servant was similarly brought up in a middle-class house and was particularly well educated in religion. She too changed after marriage took her into different physical circumstances: 'Her personal condition became of "a piece" with the wretched stone undrained hovel, with a pigsty before it, in which she had been taken. We found her with rings of dirt about her neck... In this case no moral lapse was apparent.'[61] But some sanitarians emphasised the reverse process, suggesting that bodies themselves were intrinsically dirty and the source of atmospheric and environmental contamination. 'Man poisons the air by his own exhalations', wrote one commentator.[62] Similarly, Nightingale instructed health missioners to warn the poor of the 'offensive condition of an unwashed body... the consequent poisoning of the air of the room, etc. It is the human body that pollutes the air ... The body the source of defilement of the air.'[63]

It is important to emphasise the extent to which this mutual influence was not only about the conflation of morality and physicality, but also about the interactions between bodies and spaces, between bodies and buildings. Indeed, the following passage by Benjamin Ward Richardson suggests that the significance of the atmosphere in miasmatic thought, of air being constantly breathed and exhaled, blurred the distinction between the inside and the outside of the body: that is, between the inside of the body and the inside of a dwelling; between as the medical sociologist David Armstrong has put it, anatomical and geographical space:[64]

> It has been one of the endeavours of my life to show that we living men and women make in our own corporeal structures a refined atmosphere, or ether ... an atmosphere which pure and clear brings us peace and power, and judgement and joy; an atmosphere which impure and clouded brings us unrest and weakness, and instability and misery. A physical atmosphere lying intermediate to the physical and metaphysical life ... Go into the wards of a lunatic asylum and notice amongst the most troubled there the odour of the gases and the vapours they emit by the skin and breath. That odour is from their internal atmosphere ... They have secreted the madness; they are filled with it; it exhales from them.[65]

In some conceptualisations of hospitals and homes, the building was rendered active, even anthropomorphised; for example, in the idea of a hospital which 'never rests from fouling itself' or in articles on 'Hospital Hygiene' which discussed the 'healthiness' of the buildings and enclosed space itself.[66] The reverse was also the case. The domestic metaphor was used in sanitarian discourse as a way of speaking about the body itself; the body as a home for the soul or, 'the house which the mind inhabits'.[67] Many sanitary reformers wrote of bodies as houses to be kept clean and pure. Nightingale, for example, wrote that 'the body choked and poisoned by its own waste substances might be compared to a house whence nothing was thrown away'.[68] For Nightingale, the idea of ventilating a room and cleaning the skin were part of the same hygienic principle: 'Just as it is necessary to renew the air round a sick person frequently ... so it is necessary to keep the pores of the skin free from all obstructing excretions.'[69] Or, as it was put in another sanitary pamphlet, 'pores in the skin in fact, are so many self-acting chimneys'.[70] In 'Woman as a Sanitary Reformer', Richardson wrote of the 'living house', slipping with telling ambiguity between sanitary care of the body and of the home.[71]

In sanitarian thought maintaining health, preventing disease and implementing both domestic and bodily hygiene were often constructed as a constant effort, a never-ceasing battle for order and regulation. The anthropologist Mary Douglas's central idea in *Purity and Danger*, that 'dirt is essentially disorder' and that dirt is 'matter out of place', was quite central, also, to nineteenth-century articulations.[72] In an address on public health to the National Association for the Promotion of Social Science it was written that 'man's life is a struggle and the very existence of his body is the result of an unceasing contest between destructive and repairing influences... He must never rest in his unceasing struggle to overcome the tendency he finds in everything around him to get in the wrong place. Even cleanliness, the most obvious of simple duties, is the result of the never-ending removal of impurities from places they defile to their proper receptacle.'[73] Indeed in an 1878 text the author went so far as to say that 'disorder and disease are now used as words that mean the same thing'.[74] The endless cleaning which sanitary reformers insisted upon, which the new nurses were to do in hospitals, which housewives and servants were to do in homes, and which local government authorities were to do in towns, was not only about removing dirt, but about ordering things, because disorder meant disease, disorder *was* disease.

2 Female Bodies at Work: Narratives of the 'Old' Nurse and the 'New' Nurse

Cultural, literary and visual representations of an 'old' nurse and a 'new' nurse emerged around the middle of the nineteenth century. These two figures appear again and again in discourse on health, disease and medicine. The 'old' nurse was represented as a large, elderly, disordered, working-class woman who worked without scruples regarding money-making, and without ethics or morality. The 'new' nurse was defined against this image, as young, middle-class, single (chaste), moral and pure. In countless instances, and for many decades, the old and the new nurse became particularised as Mrs Sairey Gamp from Dickens's *Martin Chuzzlewit* and Miss Florence Nightingale of the Crimea. In many ways they were equally fictitious figures. Stories about, and representations of the old nurse and the new nurse gained the currency they did because they stood for far more than a simple depiction of a change in the social, cultural and professional 'make-up' of actual nurses. They stood for a process of modernisation in the domain of health. And they serve to illustrate the ways in which women and femininity negotiated this process, and were problematically located within it.

Narratives of the old nurse and the new nurse symbolised a shift from an unregulated, disordered, unruly, undisciplined, pre-modern system to a professional, regulated, ordered, disciplined and modern one. Or more correctly, they were one version of how modernising sections of the medical/nursing world wanted to see that transition, for these figures neatly represented what was in fact a decidedly messy, difficult and incomplete process. The possibility of drawing upon an immediately recognisable image of the old nurse/midwife was arguably as crucial for the self-definition of the regulating medical profession as it was for the new reformed nurses themselves. The new nurse should also be seen as a sanitised version of the old nurse. From a morally and physically impure and filthy, ignorant, working-class hag emerged a middle-class figure of efficiency, neatness and whiteness. These figures stood for the process of sanitary reform. In this chapter, I examine the particular place and significance of the female body within sanitarian discourse: the body of the 'old' nurse, and especially that of the 'new' nurse, whose 'purity', at once moral and physical, mapped with some meaning onto ideas about disease also constructed through a conflation of the moral and the physical. The powerful cultural link drawn between middle-class femininity, female bodies and moral/physical purity, is particularly significant for any understanding of the meanings embodied and performed by the new nurses and the cultural work they did in the domain of health. In order to contextualise this discursive reading, I examine first the material and historical specifics of the mid nineteenth-century changes in nursing; the much-discussed, but surprisingly little analysed reform of nursing. The changing meanings ascribed to the figure of the nurse are not only fascinating in themselves, but are quite central to the larger cultural history of nineteenth-century medicine. Stories about the old nurse and the new nurse were about progress from pollution to purity. This was a story about rendering things clean, ordered and controlled; a Victorian narrative through and through.

RE-FORMING THE NURSE

In the mid to late nineteenth century, the bulk of nursing work in hospitals would be done by working-class women who might otherwise be upper domestic servants.[1] The process of change took two broad forms. First, most significant changes were brought about through the re-forming of these working-class women within a culture of middle-class values, usefully listed by the committee of the Nightingale Fund as sobriety, honesty, truthfulness, trustworthiness, punctuality, quietness and orderliness, cleanliness and neatness.[2] They were culturally recoded, if you like, as middle-class. Second, major changes were brought about by the management of working-class nurses by middle- and upper-class women, increasingly gaining positions as 'sisters' or 'head-nurses' of the wards in the large London voluntary hospitals and as matrons or lady superintendents of institutions throughout Britain and soon after, colonies in South Africa, Australia and Canada. This moral management of working-class women was also implemented in the numerous religious sisterhoods and deaconesses' institutions which were established from around the 1840s: for example, Elizabeth Fry's Quaker Institution for Nursing Sisters; the Sisterhood of the Holy Cross and the Sisterhood of Mercy, largely under the influence of Pusey and the high-church Tractarians; the St John's House Sisterhood which undertook all the nursing and household responsibilities of King's College Hospital, London.[3] To a greater or lesser extent, all the new nursing sisterhoods and hospital hierarchies recognised so-called lady-nurses or lady-probationers as distinct from ordinary nurses. Rules of the St John's Sisterhood in 1867 stated that there were three categories of women in the institution: '1. Sisters: Ladies who are willing to devote themselves to the work of attending the sick and poor in Hospitals or elsewhere, and of educating others for the like duties. 2. Probationers: Women under training as Nurses of St John's House. 3. Nurses: Women who have passed satisfactorily their period of probation.'[4] Nurses and probationers received a salary of £15 for undertaking domestic duties in the House itself as well as the duties on the wards of the Hospitals and in private homes. The sisters accepted no salary, paid for board and lodging at the House, and supervised the work of the nurses.[5] As Martha Vicinus has noted, the 'employment' of such women did not 'disturb middle-class prejudices about ladies' working for money'.[6]

The first probationers which the Nightingale Fund Committee employed in 1860 were working-class women. Assistance was given to them in reading and writing, and many were dismissed or withdrew, some because of drunkenness, some because of sickness.[7] After some time, the Nightingale Fund shifted its focus to include better educated women, and by the late 1860s also recognised three broad categories. Its 'special probationers' paid £30 to the Fund, and received no salary. It also announced 'occasional vacancies...for the admission of gentlewomen free of expense, together with in some cases, a small salary during the year of training'.[8] Ordinary probationers received a small salary, a gratuity on completing each year's work, as well as board, lodging, washing and clothing maintained at the expense of the Fund.[9] Like the sisters at St John's House, the special probationers were prepared for superior positions as matrons, lady superintendents or sisters of wards in large hospitals. The ordinary probationers would be under their management and their 'moral control'.[10] In secular hospitals, as in the sisterhoods and deaconesses' institutions, the relationship between differently classed women was largely one of mistress and servant, a relationship described thus: 'It is the servants of a house whose properly directed work has to keep it clean and sweet. It is the lady of the house only who can regulate that labour, and by adding to it her own intelligent supervision can make it effective for cleanliness, and for the health to which that cleanliness is the first and greatest step.'[11] The construction of domestic woman as an 'indoor' sanitary reformer made the model of a female-managed household particularly apt for hospitals.

While there were certainly significant changes in the domain of nursing, there was no real consensus on what class of women nurses should be drawn from, how and what they should be taught, what responsibilities they should hold, how much they should be paid, whether religious nursing sisterhoods were appropriate or whether ladies' voluntary work in hospitals was acceptable or useful. All of these issues were under question in a general atmosphere of reform.[12]

There has been a tendency for historians to locate reforms in mid-nineteenth-century nursing within an overall framework of changing medical knowledge, narrowly conceived.[13] Around the middle of the century, it is argued, doctors began to demand more reliable and efficient assistants to observe, report on and treat their patients, in order to assess effectively new medical and surgical techniques. Certainly there were significant developments in medical practice at this time, particularly in surgery, and the whole area of hospital mortality and morbidity rates and statistics was becoming important. Hospital governors and doctors looked to all areas of the treatment of patients for improvement, including nursing. Yet, overall, there has been a fairly systematic privileging of such medical factors, and a disinclination to locate the changes within a broader politics of health. This narrow interpretation fails to explain the central involvement of middle-class women and of religious women in the changes. It fails also to explain the considerable authority such women came to wield in institutions, as well as the opposition to them which arose in certain medical quarters. Moreover, the fact that these changes were specifically about the health and ill-health of the working class has not been accorded much significance. This point is quite crucial to an understanding of the nature of middle-class women's place in institutional health care at this time. It seems to me that rather than thinking of nursing reform as part of local changes in medical therapeutics, it needs to be reassessed as operating well within that cluster of developments I isolated in the last chapter as a new 'politics of health' – public health, sanitary reform, sanitary science and social science. More specifically, nursing reform was a peculiarly feminine intervention into this politics – part of middle-class women's philanthropic culture.

The mid nineteenth-century reforms in nursing took place quite specifically around the question of the moral/physical health and hygiene of the working class. The changes occurred in several locations: in the large voluntary hospitals (which admitted only working-class people); in the workhouse infirmaries which the New Poor Law had produced; and within the homes of the urban poor, as extensions of the religious and philanthropic 'visiting' discussed in the last chapter. The question of nursing was discussed as part of the new social science, and specifically constituted as a social–scientific 'problem' by the public health section of National Association for the Promotion of Social Science.[14] It was argued quite simply that improved nursing would improve the health of the poor. In quite material terms it probably did, given the centrality of the concept of cleanliness to reformed nursing. A reduction in fatal post-surgical infecton can plausibly be attributed to changes in nursing, for example. Yet this 'improvement' in nursing enabled a whole range of processes important to the new sanitary science and to a politicisation and bureaucratisation of health and hygiene. What constituted 'nursing' as problematical in the mid-nineteenth century, and what presented it as a subject for 'social science', was largely the same set of issues which constituted hospitals, health, and the moral and physical bodies of the working class as problematical, as discussed in Chapter 1. The increased order, regulation, accountability, and basic cleaning which came with the new nurses allowed a more specific and reliable classification of bodies and diseases, more accurate statistics on births and deaths and a systematic observation and discipline of working-class behaviour in a range of locations. It is no coincidence that Florence Nightingale established herself as both a premier authority on the reform of nursing and the management of hospitals on the one hand, and as a statistician of disease and the health of populations on the other. New incitements to hygiene and health, new technologies of population maintenance and surveillance, newly disciplined bodies, I would argue, are more fruitful ways in which to frame and understand the massive changes that occurred in nursing in this period. As Mitchell Dean and Gail Bolton have also argued, the reforms were about an 'administration of poverty'.[15]

Foucault wrote of new cultural incentives towards 'the healthy, clean, fit body; a purified, cleansed, aerated domestic space; the medically optimal siting of individuals, places, beds, utensils... Doctors will, moreover, have the task of teaching individuals the basic rules of hygiene.'[16] While he saw all of this as the practical domain of medical men in eighteenth-century France, I see it in nineteenth-century England as being unmistakably the practical domain of women in hospitals and urban spaces. Like the female philanthropists examined in Chapter 1, the new nurses were the central agents of a marked new moral dimension at play within charitable institutions and in urban and domestic spaces; their role was configured through the gender/class politics of sanitary reform. For example, one exponent of 'Hospital Training for Ladies' wrote that '[a]nything that brings the well-born, educated women into practical contact with the very poor, is an equal advantage to both... The lady would acquire a better understanding of the pauper; the pauper would be humanized by the lady.'[17] The appointment of gentlewomen such as Agnes Jones to the Liverpool Workhouse Infirmary is best understood as an extension of women's philanthropic and domestic role in the arena of health and hygiene. And it is especially significant that such women not only worked, but lived in these institutions. They provided, in Anna Jameson's words, a 'domestic, permanent, and ever-present influence, not occasional inspection'.[18] They assumed a role which was quite explicitly about reform, of both working-class nurses and working-class patients, and in moral, physical and sanitary senses. From the basis of such strong convictions about the reforming and humanising power of their moral and spiritual influence, women gained an authority to undertake sometimes very major material changes within institutions. The management of charitable institutions was one public area in which it was possible for women to gain an expertise which politicians and administrators sometimes chose to recognise. The advice of women such as Florence Nightingale, Louisa Twining and a host of others, was heeded on a whole range of issues which concerned prisons, hospitals and asylums, from architecture to overcrowding to the suitability of particular staff. It was this type of expertise, rather than specific medical or nursing knowledge, which determined the nature of middle-class women's managing roles in institutions. Often enough, such women claimed and wielded an authority far beyond that expected by, and acceptable to, governors and doctors. Major disruptions occurred in the 1870s, 1880s and later between new lady superintendents and medical hierarchies. The 'crisis' at Guy's Hospital in 1879 and 1880, for example, was a quite public affair, as was the contention for power between the St John's House Sisterhood and the governors of King's College Hospital.[19] As with Nightingale in the Crimea, however, women's potentially disruptive and challenging work in institutions was often reconstituted (and diminished) in public discourse as tendering emotionally to individual sick men, by placing hands on brows, carrying lamps and so on.

The women's sanitary organisations discussed in the last chapter and the reform of nursing operated well within the same discourse. And not surprisingly, the material connections between the two were also quite direct. The Ladies' Sanitary Association saw the establishment of a 'Training Institute for Nurses' as 'one of the most important points of the Association's proposed work'.[20] The secretary, Mary Anne Baines, responded to Nightingale's pamphlet 'Nursing, What it is and what it is not', with another titled 'Sanitary Reform, What it Is and What it Is Not'. She wanted some of the money collected for the Nightingale Fund to be allocated to the Ladies' Sanitary Association which would carry into effect the planned training institute for nurses which the Nightingale Fund had (as yet) failed to do. Baines wrote to Nightingale:

> permit me to say, that the value of the Ladies Sanitary Association can in no degree be lessened in the mind of the public, when the fact is known that its leading intention is to preserve *Health*, rather than to tend *Sickness*. *Hospitals for the Sick* abound and are well supported, in and beyond the Metropolis, these possess ample machinery for the instruction of nurses in every department appropriate... Philanthropy, however may take a new and important direction, by establishing *Hospitals of Health*... [I]n furtherance of that great work which is the glory of the age – the work of *Sanitary Reform*.

'Impudent woman', Nightingale wrote in the margins of this letter.[21] By the end of the century, nurses, especially district nurses were still being articulated specifically within a sanitarian discourse. For example, one district nursing organisation discussing the knowledge which should be imparted to nurses, argued that 'foremost among these is the subject of hygiene for the District Nurse should above all things be a Sanitary Reformer, and practical teacher of health matters in the Homes of the Poor. It would most likely be desirable that these Lectures should deal with such subjects as ventilation – drainage – water supply – nuisances – and how to deal with them and other sanitary matters.'[22] Moreover, the mid nineteenth-century connections between such knowledge and religious instruction were still quite visible in the early twentieth century.[23] The question of nursing, sanitary reform, the management of charitable institutions, indeed the whole project of 'health' as opposed to sickness, resided as much in the hands and culture of women's philanthropy, as it did in male medical culture.

THE OLD AND THE NEW NURSE

All of these changes in nursing were schematised into the story of old and new nurses, told and retold as a narrative of sanitary reform. A most explicit visual example of the old and new nurse is an illustration from an 1888 issue of the *Nursing Record*, a British journal which prided itself on its independence from medical control and which included feminist columns on women's work, the suffrage campaign and so on (see Figure 2.1). In this illustration, the two portraits, captioned 'Fifty Years, 1838, 1888, Then, Now', show the 1838 nurse as middle-aged, working-class, with an alcohol bottle as her symbol. In countless other such representations, the old nurse was similarly constructed as literally aged and often as a widow, as personally disarrayed, as drunken and ignorant. Her immorality was sometimes represented sexually, but more often in terms of insobriety, disorder and personal dirtiness. In the midwife version she was often a self-interested and unscrupulous businesswoman as well. The old nurse was the more straightforward and unambiguous image of the two nurses – a stable figure, always depicted more or less the same way, and always carrying meanings of dirt and disorder. The 1888 nurse in this image is young and neat, confidently confronting the viewer's gaze and most importantly, the bottle has been replaced by a cross. The new nurse was typically represented as middle class, personally clean and morally pure, with a discernible, but checked religious spirituality. Far from wishing to profit from her work, her overriding characteristic was self-sacrifice, but occasionally she was depicted as a 'new woman' for whom career and financial independence were central. This figure became the frontispiece for subsequent issues of the *Nursing Record*. It captured the way in which the particular group of 'new nurses' editing the journal wished to be represented. The corpulent excess of the old was succeeded by the trim efficiency of the new. Of course, all of these versions of the nurse worked through an already established repertoire of stereotypical femininities, but they were also images which carried particular significance within sanitarian discourse.

Figure 2.1 The Old Nurse and the New Nurse: 'Fifty Years'
Source: *The Nursing Record and Hospital World*, 20 December 1888.

Anne Summers has examined the figure of the 'old' nurse as representative of early nineteenth-century independent practitioners. She argues that such practitioners 'offended a growing and many-stranded movement for nursing reform...Male physicians and surgeons, religious reformers of both sexes, and all those anxious to expand professional opportunities for women, combined both consciously and unconsciously to deny Mrs Gamp and her ilk a respected place in the provision of care for the sick.'[24] While this occupational rivalry was certainly important, the specific way in which the 'old' and 'new' nurses were constructed locates them firmly within a discourse of sanitary reform. The figures gained the currency they did, precisely because they encapsulated the fundamental elements of the sanitary campaign: the reform of working-class to middle-class standards; morality over immorality; cleanliness and purity over dirtiness; youth and progress over age; order over disorder; knowledge over ignorance. This was a story of sanitary progress. The new nurses were reformed and sanitised versions of the old nurses, and in turn, they functioned to reform and sanitise working-class patients and charitable institutions as a whole.

The narrative of the transition from the unclean, insanitary old nurse to the morally and bodily pure new nurse maintained a remarkable resonance over time and place and recurred constantly in late nineteenth and early twentieth-century discourse on health and disease. In one article, the old nurse was depicted as 'slovenly' and 'ignorant'; she displayed a 'positive cruelty', a 'wilful neglect', 'inattention, uncleanness...drunkenness'.[25] In another, the old nurse was described in animalistic and evolutionary terms as 'coarse-faced, thick of limb, heavy of foot, brutal of speech; crawling up and down the stairs, or about the wards'. By contrast the new nurses of modern hospitals were the perfection of sanitary principles: 'Go now through the corridors and wards of a modern hospital – every nurse you meet will be neat and trim, with her spotless dress and cap and apron moving quietly but quickly to and fro, doing her work with kindness and intelligence.'[26] An obstetrician wrote: 'she takes care that not the slightest trace of disorder shall appear in her dress or on her snow-white collar and cuffs'.[27] In 1887, some 'new' nurses gave the moral distinction between themselves and the old nurse bodily expression:

> There was no need of a straight-jacket when she [the old nurse] was there; with her, moral suasion had a secondary place, and physical force was her great virtue...The science of medicine has advanced,

with it the art of nursing, so that the figure of old would be as much out of place in the modern sick-room as a bull in a china shop. See her now in her snow-white cap and spotless apron, gliding almost imperceptibly from bed to bed... her very presence sending a stream of light and cheer into a roughened soul.[28]

In such depicitions, if the old represented brute physicality, the new represented ethereal spirituality and purity.

In hospitals and homes, it was the nurse who took major responsibility for cleaning and sanitising spaces and bodies. 'The fear of dirt is the beginning of good nursing,' wrote Nightingale.[29] Significantly, the sanitising and purifying of their own body was often seen to be the first step in this process; exquisite personal hygiene was the work and the mark of the new nurse. The efficiency of the nurses in the City of Hygeia, wrote Benjamin Ward Richardson, stemmed from 'the care with which they attend to their own personal cleanliness'.[30] A manual for midwives instructed that 'neatness and cleanliness characterised not only her person and dress, but the entire sphere of her duties'.[31] The purpose of this neatness and ordering was far from aesthetic. Order was seen to be quite literally a sanitising, if not a therapeutic, mechanism. The feminist Bessie Rayner Parkes wrote thus of a nurses' training school in 1865: 'In their own home, all is excellently arranged; neat and airy bed-rooms, good food well served, appropriate dress – everything to counterbalance the miasma of disease.'[32] Miasmatic and sanitarian discourse dictated very specific work practices. Because miasmatic theories held that the human body constantly ridded itself of impurities, anything which came into contact with sick bodies was considered polluting; the work was constant and endless. The appeals for nurses to clean were almost always couched simultaneously in terms of medical therapeutics and moral urgency.

When middle-class lady superintendents entered various charitable institutions newly under their charge, the process of reforming and 'purifying' not only included, but usually began with, the bodies of the women who worked there. For example, Lucy Osburn, sent by the Nightingale Fund to manage and reform the Sydney Infirmary, wrote to Nightingale in 1868 of the

> dirty 'frowsy' looking old women, slatternly untidy young ones all greasy with their hair down their backs with ragged stuff dresses... I think I told you they slept in the Wards behind screens & that I got a dormitory assigned to them in which we got 4–posted

iron bedsteads wash handstands & looking glasses the Sisters taught them to do their own hair there was no excuse for being dirty when they had a basin to wash in.[33]

Sanitary ideas of dirt and cleanliness, disorder and order were precisely the terms in which English women at the Sydney Infirmary constructed their difference from the colonial women they encountered. They placed themselves within, and reproduced the narrative of the old nurse and the new nurse. Mary Barker had worked at St Thomas's Hospital in London for eight years and clearly saw herself as a 'new' nurse despite her working-class status. Indeed, Barker was an exemplary model of a working-class woman successfully reformed and coded with all the middle-class values of a new nurse. In a letter to Nightingale she portrayed the colonial women at some length, shifting in true sanitarian manner, between their personal disarray, the ill-health of the patients and the hygiene of the wards:

The Wards was [sic] in a very rough dirty state...there was [sic] dirty old gowns, skirts and shawls hanging all round the beds... and the Patients looking so miserably dirty, I must not say anything about the ventilation. The Nurses were dressed in all colours with old jackets and old gown skirts in rags all round the bottom and the largest crinolines I ever saw no caps and some not a bit of apron, some with their hair in Chignons some with it combed of [sic] ther [sic] faces, others hanging down over shoulders in all cases looking as if it had not been combed for a week. I think the scrubbers at St Thomas Hospital were a respectable class of women in comparison with what we found as day Nurses at the Sydney Infirmary... women who consider themselves good Nurses, would let these patients lay in there [sic] beds unmade for weeks and not even wash there [sic] hands and face...They are so lazy, that when they se [sic] a patient covered with bed sores and vermin, it is looked upon quite as a matter of course in the colony.[34]

The sanitary reform of the hospital began here, with the material reforming, correcting and disciplining of the female nurses and servants under the control of new lady superintendents. Newly groomed, uniformed, cleaned and regulated, they were to be the foundation from which the moral, physical and sanitary transformation of the institution as a whole was to occur. Even as the broad public health project aimed to sanitise working-class bodies and homes, so it was seen to be possible to reform and sanitise these nurses and thereby the hospital in which they lived.

Yet it is interesting to note that the male nurses who actually dominated colonial hospitals (there, the 'reform' of nursing was about 'feminising' nursing) were seen to be part of the insanitary condition of the hospital which had to be replaced, rather than reformed. Their wards were described as 'grimed and insect infested... dirty corners where all kinds of filth and rubbish was stowed away... sufficiently attested to what their ideas of order and cleanliness were... I hope they will never again resume their work where women have been introduced.'[35] While working-class femininity could be manipulated into meanings congruent with respectability, cleanliness, and order, working-class masculinity could not. Put simply, purity and masculinity did not fit together. In the discourse of sanitary reform, as will be seen, women could embody purity *and* pollution in a discursively powerful way. In part, this is the reason why male nurses were omitted entirely in the telling and retelling of the sanitarian narrative of old and new nurses, notwithstanding the presence of male nurses in many colonial hospitals, in some American institutions and in English military hospitals. The story of old and new nurses resonated far more deeply around issues of purity and pollution if constructed in terms of a shift between types of women, types of female bodies.

Miasmatic metaphors of lightness and darkness, bad smells and fresh air, were constantly used to tell the story of the old nurses and the new lady nurses in hospitals. An Australian woman wrote that hospitals in England and the colonies were being reformed by 'the simple presence of gentlewomen; where an atmosphere of kindness and gentleness, truth, justice and courtesy is breathed'.[36] The pure thought of gentlewomen in institutions, she wrote elsewhere, would 'fall like sunshine upon uncleanness, but keeps its own purity the while'.[37] A slightly later article suggests the extent to which the sanitarian language of purification and atmospheres persisted, even when a different language of germs, sterility and antisepsis was available: 'I can bear testimony to the great civilising influence they have been, acting like a moral antiseptic purifying the whole atmosphere.'[38] There was an image of ladies and gentlewomen needing only to be present in hospitals for the most important part of their work to be achieved, for hospitals and working-class homes to be rendered more pure and moral: it was said that the presence of gentlewomen 'introduced a refined and home-like influence which is said to be, in itself, curative'.[39] Order, purity, respectability would simply emanate or in the more commonly used language 'exhale' from such women. Within the context of hospitals and sanitary reform, and in sanitarian discourse, that purity which 'exhaled' from a lady assumed a therapeutic significance. Additionally, the physical work involved in nursing was seen to be beneficial, if not actually therapeutic for these women themselves. According to one commentator, lack of work, seclusion and introspection were 'as certainly dangerous to spiritual health, as the miasma of standing water to health of the body'.[40] Picking up on such popular perceptions of well-off and idle spinsters, the doctor Elizabeth Garrett, although not a supporter in general of ladies' voluntary work in hospitals, saw it as 'positively invigorating':

> Constant exercise in large and airy wards, employment of the kind which prevents morbid introspection or continuous mental exertion, absence of anxiety, regular and early hours, simple diet, and a life at least much less dull than that of most single women, combine to form a sum of conditions under which the health of most ladies would rapidly improve.[41]

For Eva Lückes, matron of the London Hospital, even the relationships between nurses within institutions were best explained with miasmatic metaphors:

> Think of the enormous power you exercise over each other by your daily example... It is, so to speak, the mental atmosphere with which we are surrounded, and as we were speaking just now of the air we breathe, and the effect which its condition produces upon us, and the effect which every *one* produces upon its condition, is it not a matter of vital importance to keep this mental atmosphere pure and invigorating? Do you see what a difference it makes, metaphorically speaking, whether you are contributing your share of oxygen, or more than your share of carbonic acid?[42]

PRECARIOUS PURITY

In this discursive universe, order itself meant cleanliness. Nineteenth-century nurses might be seen as a crystallisation, a literal embodiment of such a statement. They enacted not only Mary Douglas's observation that 'dirt is essentially disorder... dirt offends against order' but also her insistence that rituals of bodily cleansing and purifying serve as metaphors for larger social ordering. Her early articulation of the body as an intensely symbolic structure is important in this context:

> The body is a complex structure. The functions of its different parts and their relation afford a source of symbols for other complex structures. We cannot possibly interpret rituals concerning excreta, breast milk, saliva and the rest unless we are prepared to see in the body a symbol of society, and to see the powers and dangers credited to social structure reproduced in small on the human body.[43]

The culturally and discursively constituted body was a major site on and through which ideas about purity and pollution, cleanliness and dirt were produced and made meaningful.

Of course, there has been a vast amount of largely feminist writing about the body since Douglas's early work. As I discussed in the Introduction, there has also been considerable feminist historical interest in the female body as the object of medical science and of medical practice.[44] Yet there has been minimal concern in this literature with the cultural uses of the female body when it is positioned not as patient but as practitioner. What happens when we turn to read the body of 'woman' not as object but as subject of medical and health practices? What does the specifically female body of the nurse, old and new, mean within a culture based on 'morally resonant polarities' and within concepts of purity and pollution? One thing is certain, the body of woman-as-practitioner means very different things to the body of man-as-practitioner. This is often not grasped by writers and theorists who often think of 'the body' as if it came in one generic type. As Ludmilla Jordanova writes, it is often not male bodies, but female bodies which 'permit cultural investment'.[45] The specificities of the *female* embodiment of nurses need particular attention.

As we have seen in Chapter 1, the domain of health, disease and illness worked through deeply embedded concepts of purity and pollution, understood at once morally and physically. And the resonance with which the narrative of the old and new nurse mapped onto this language is striking indeed. It did so, less as a result of some particularity of nursing itself, than as a result of the particularities of 'woman' which nurses embodied. In that 'nurse' came to be closely, almost identically aligned with 'woman' over the nineteenth century, the cultural construction of one always directly effected the construction of the other. If sanitary reform was conceptualised in 'morally resonant polarities', in terms of dichotomies of filth and purity, then so was the representation of 'woman' and so was the representation of 'nurse'. Moreover, and this is crucial to an analysis of the way in which the female body worked discursively within the domain of health and disease, women's purity and impurity were expressed at once morally and physically.

There is a substantial feminist literature on the deeply binarised construction of 'woman' as good and evil in the nineteenth century, on the significance of the related images of morally/physically impure prostitutes and morally/physically pure domestic woman.[46] The conviction that there was a connection between immorality and bodily disease found its ultimate nineteenth-century expression within the domain of female sexuality and through the representation of female bodies. Male bodies never carried these meanings with such intensity, except perhaps within the context of religious purity and asceticism. For much of the nineteenth century, prostitutes were seen to be the source of venereal disease, which arose from their bodies *de novo*; they were often represented as the disease itself.[47] Urban areas were to be 'sanitised' of prostitutes as it was sanitised with sewers, and a connection between the body of the prostitute and sewers and drains was not uncommon.[48] The moral/physical filthiness of prostitutes was an immensely powerful image, which mirrored the moral/physical purity and cleanliness of chaste women. Leonore Davidoff and Catherine Hall have identified the importance of a concept of bodily purity for women of the new middle class, and the central linking of physical and moral purity. They write:

> All serious Christians were exhorted to turn their backs on 'the world'...The experience of conversion was often described as a process of cleansing and purification from the pollutions of the world...The concept of purity had taken on a special resonance for women partly because of fears associated with the polluting powers of sexuality. One of the distinguishing characteristics of the middle class was their concern with decorum in bodily functions and cleanliness of person. Thus, maintaining purity and cleanliness was both a religious goal and a practical task for women.[49]

The significance of the new lady-nurses and lady superintendents entering hospital work from around the 1860s should be read in this light. There was a highly successful bid to link the new nurses with cleanliness, maximising the potential of that 'purity' inherent in the cultural construction of middle-class women. This was a concept of cleanliness wrapped up in moral terms and which could be effectively secured in hospitals by exercising women's domestic authority and skills. These women embodied concepts of cleanliness and morality which were consistent with the objectives of sanitary reform and with disease theories which saw moral and physical purity as one and the same. In the same way that nineteenth-century sanitary reformers were convinced of the material effect of immoral and impure bodies in defiling the atmospheres of institutions and homes, so moral and pure bodies – chaste female bodies – played a particular role in bringing about order and cleanliness. Young, chaste, middle-class, they both embodied and personally demonstrated moral and bodily cleanliness, as well as becoming the agents through which institutions were to be materially cleaned and ordered. The reform of nursing within hospitals was very much part of a cleansing and purifying as well as a domesticating process, both material and cultural. Bodily hygiene and domestic hygiene, cleaning one's self and cleaning a room, were by no means distinct practices, but were rituals discursively intertwined in very powerful ways.

The resonance between the construction of 'woman' and sanitarian discourse was not simply about a similar cultural dichotomy of purity and pollution being in place. It was also about the nature of women's 'purity'. For no matter how pure the woman, this was a precarious state. There was a sense in which, like efforts to sanitise a hospital which 'never rests from fouling itself',[50] women's bodies always contained a threat of pollution. There was always a potential contamination or impurity specific to female bodies and to feminine moral existences. Dominant medical understandings of menstruation, not to mention longstanding religio-cultural understandings, produced a notion of women's bodies as always polluted. As Deborah Lupton has written: 'The potent symbolic meanings of blood, relating to death, pain, loss of control and warfare – that is, general bodily and societal disorder – and the regular emergence of menstrual blood from the uterus and vagina – parts of the body which are themselves considered dirty and contaminating – combine to render menstrual blood a highly meaningful and anxiety-provoking fluid.'[51] The female body itself was 'matter out of place'. For much of the nineteenth century especially, women were constructed as subject to an internal pollution, produced *de novo*, as it were. In this sense, woman always *needed* purifying. In her moral life also, nineteenth-century woman was constructed as always having a capacity to 'fall'; her moral purity was never secure either.[52] If chaste female bodies were the site on which sanitary codes of order, purity and cleanliness were written, there was always a sense in which this purity was under threat, both from outside the body in the form of filth (and here the intermittent representation of disease as male sexuality, as a 'detestable impregnation', is significant),[53] but also from the inside the most chaste female body.

A cultural ambivalence about the female body runs through the representations of the old and new nurse. While they might be snow-white and spotless, the most pure female body could also be seen as inherently diseased. Women embodied the power of contamination. This instability in the representation of the female body did not undermine the usefulness of nurses as symbols of sanitary reform. Rather, this was precisely why they represented the concept of sanitary cleanliness so effectively. The concept of 'sanitary' invokes a state of cleanliness always defined through the filthiness and impurity it displaces, and which is continually threatening to recur. Thus the discourse of sanitary reform could be mobilised at once around the purity of the new nurses and the filthiness of old nurses. And it could be mobilised around the purity/pollution embodied by any one woman. It was precisely *this* metaphoricity of the female body which lent so well to concepts of 'hygiene' and 'sanitary'.

Current usage of 'sanitary' has crystallised the nineteenth-century usage, which always worked through a simultaneous sense of the pure and impure. 'Sanitary' is only employed in the late twentieth century in connection with the purification of that constructed as fundamentally dirty: toilets, sewage and female bodies. A contemporary dictionary definition exemplifies this perfectly:

> **sanitary**: of the conditions that affect health esp. with regard to dirt and infection... – *engineer*, dealing with systems needed to maintain public health; – *napkin*, absorbent towel used during menstruation; – *ware*, porcelain for lavatories etc.[54]

The concepts of 'hygiene' and 'sanitary' have retained their connection with female bodies deeply constructed as contaminated and contaminating. Occasionally, however, this close connection between the various meanings of 'sanitary' were made quite explicitly in early texts. In a section on how to manage nurses' homes, the authors of the book *The Modern Hospital* moved from discussion of a contaminating female sexuality, to plumbing and ventilation, to menstruation virtually without pause:

One morbid pupil, with erotic tendencies, can sometimes pollute a whole training-school. Not only should the health of the girls individually be looked after carefully... but the health of the home, as a whole, should be regulated by proper hygienic and sanitary equipment of every sort... It is not enough that the plumbing and ventilation shall be right, and that the pupils shall be compelled to use fresh air in large quantities... [but that] they should be taught to keep their bowels open, and the supervisor should see that they do so. They should be taught to protect themselves before and during their menstrual periods, and they should be supplied with proper sanitary appliances for use at that time.[55]

Given the Victorian anxiety about boundary order, the fact that the ordering and cleansing of hospitals began with and through female bodies is significant indeed. For the female body is one in which boundaries are perceived as insecure and uncertain. Elizabeth Grosz writes of 'a broadly common coding of the female body as a body which leaks, which bleeds, which is at the mercy of hormonal and reproductive functions'.[56] If any culture ascribed such meanings to the female body, it was Victorian culture. It is partly for this reason that the figures of the old nurse and the new nurse were so super-invested with meaning. The new nurse represented a momentary control of this 'uncontrollability', a precarious purification of the contaminated female body, an attempt to order that which is messy and unbounded. In the domain of public health and medicine, nothing symbolised the process of purifying, sanitising and moralising more evocatively than the partly-fictional, partly-real figures of the 'old' nurse and the 'new' nurse. The nurse was a central player in Victorian obsessions with dirt, pollution, impurity, control and order.

3 'Disciplines of the Flesh': Sexuality, Religion and the Modern Nurse

'Discipline' was a concept and a word commonly employed during the nineteenth-century reforms in the administration of institutional health. In this modernising process, doctors, matrons and managers argued over changing modes of organisation of patients, medical students and probationer nurses. The discipline of nurses, it can be argued, was one of the major expressions of this reform and modernisation. But nurses were both agents and objects of institutional discipline. Their own bodies were controlled, disciplined and trained, even as they were the agents through which the bodies of patients and inferior nurses were trained in various ways. The multitude of nineteenth- and early twentieth-century articles about nursing which deal with precisely this, with titles such as 'Discipline and Etiquette', 'Hospital Discipline and Ethics', almost demand to be read against Foucault's *Discipline and Punish*.[1]

Employing Foucault in order to read 'discipline' in hospitals locates analysis of nursing within sociological debates about modernisation, secularisation and rationalisation. The modernisation of Western culture has largely been about its secularisation. And this secularisation has been effected through, and reflected in, the shift in the control of bodies from religion to medicine – a transfer of authority and power between what the sociologist of medicine Bryan Turner has called two 'disciplines of the flesh'.[2] As Turner writes, following Foucault, modernisation has been about 'the transfer of moral regulation from the church to the clinic...medicine occupies the social space left by the erosion of religion'. And in a later publication, he writes that 'medical belief and practice operate as a quasi-religious institution'.[3] Nursing was clearly implicated in this process of the medicalisation of bodies and the rationalisation of institutions, but in no simple way. The domain of nursing occupied an ambiguous, highly flexible and quite unique discursive location between religion and rational science. And it is this transitional location where the significance of nursing in this scheme of modernisation lies. If 'secularised' medicine became a 'quasi-religious institution', then it can be argued that what produced this 'quasi-religious' meaning most effectively in the domain of the health care/control of bodies in the nineteenth century was modern nursing; specifically the figure of the 'new' nurse. In this chapter I offer such a large reading of modern nursing, exploring the connection between religion, discipline and sexuality, as they pertained to questions of modernisation and secularisation.

There are several arguments to be made about nursing and the broadly defined processes of modernisation. First, I render more complex the common interpretation of the significance of sexuality in nineteenth-century nursing by analysing it within the framework of the discipline and ascetic control of bodies. The rituals and work cultures of a range of modernising institutions were, as Turner has suggested, 'anticipated by the discipline of the monastery in which bodies were subordinated to ascetic rules of practice'.[4] 'Ascetic', meaning to be severely abstinent and austere for some spiritual benefit, also carries a strong sense in which this is to be achieved through self-discipline, and necessarily entails some hardship, sacrifice, even pain, and is clearly one mode of discipline. For modern nurses, like the nuns and religious women from whom so much about their work and identity derived, sacrifice – of self, money, leisure, bodily comfort, marriage, motherhood – was a central organising principle. What needed sacrificing, but also controlling and disciplining most in a culture of religious asceticism, as in reformed hospitals, was sexuality.

The standard argument about nineteenth-century nursing is that efforts to control sexuality as part of hospital reform were primarily about imposing a culture of respectability on the occupation to 'raise the standard' and enhance its attractiveness and viability for single middle-class women.[5] While this interpretation certainly retains some validity, the significance of controlling sexuality has far more extensive implications. It was a fundamental way in which pre-modern disorder was rationalised in the context of 'asceticism' and modernisation. Yet, even as female sexuality could be sacrificed and controlled in this process, it could not be erased. Rather, female sexuality always retained a submerged and threatening presence in nineteenth-century bourgeois Western culture generally, as well as in the specific context of nursing and hospitals. A second objective of this chapter is to demonstrate how concepts and experiences such as 'sacrifice', 'sexuality' and the ascetic control of bodies all have a crucial gendered dimension, which needs to be brought into play with sociological analyses of modernisation and secularisation. For example, the 'sacrifice' demanded of nurses certainly drew from religious discourse, but was also a particularly intense version of the sacrifice of self asked of women in general in this period.

A final argument which I develop in this chapter deals specifically with Turner's analysis of religion and nineteenth-century nursing. In his essay, 'Weber on Medicine and Religion', Turner takes nursing as exemplary of Weber's theories on the transformation of religious to secular culture.[6] This theory suggests that religious calling or 'vocation' became the model for professions, which subsequently came to be defined by market-forces, technical, mechanical and rational logic, and that Protestantism in particular, brought the culture of monastic asceticism into everyday life. Weber wrote: '[The] uniform goal of this asceticism was the disciplining and methodical organization of conduct ... and its unique result was the rational organization of social relationships.'[7] According to Turner, nineteenth- and early twentieth-century nursing displayed a classic Weberian pattern of a secular calling with religious roots, which over time came to have no need for religious legitimation, and which was ultimately driven by 'predominantly market interests of social closure and occupational power'.[8] He identifies a secularising trend in nursing which is broadly correct, yet is far too straightforward. Turner himself writes of secularisation as an 'uneven and contradictory process of cultural change', yet often fails to identify the nature and implications of these complexities, especially when they centre on questions of gender.[9] Here, I essentially reverse his approach to the connection between nursing and religion. That is to say, while Turner draws meaning from the process of secularisation and bureaucratisation, I want to identify and draw meaning from the *ongoing* religious legitimation evident in nursing, and therefore within the domain of health care as a whole. I do not wish simply to interpret this as some sort of 'residual' set of values, but as a situation which was fundamentally tied up with the femininity of new nurses, for whom being 'rational' and 'secular' held specific difficulties. The point which needs making is less one about secularisation, than about the almost permanent ambiguity in the discursive positioning of nursing between religious and secular/scientific fields of meaning. It is precisely this ambiguity which did some important work in the modernising politics of health, in a sense facilitating or 'allowing' the transition of medicine to a secular domain.

TRAINING AND DISCIPLINE

The idea of nurses' 'training' which emerged in the mid to late nineteenth century is commonly understood to refer to their system of education. However, in a Foucauldian context, the 'training' of nurses takes on a different meaning which in many ways illuminates far more keenly the dynamics at work in modernising hospitals. Nurses were trained in behaviours, relationships, modes of surveillance of their patients and each other, as much as they were trained in elementary anatomy and physiology, for example. In Foucault's usage, a disciplinary regime 'produces subjected and practised bodies, "docile" bodies. Discipline increases the forces of the body (in economic terms of utility) and diminishes these same forces (in political terms of obedience)'.[10] Foucault argued that discipline imposes upon subjects a relation of 'docility–utility' through a whole range of practices, most of which resonate clearly with the modernisation of nursing: the incessant regulation of detail; the imposition of timetables which structured the constantly repetitive cycles of work; precision of command; systems of 'micro-penalties'; hierarchical observation and examination. For example, he lists micro-penalties of 'time (lateness, absences, interruptions of tasks), of activity (inattention, negligence, lack of zeal), of behaviour (impoliteness, disobedience), of speech (idle chatter, insolence), of the body ('incorrect' attitudes, irregular gestures, lack of cleanliness), of sexuality (impurity, indecency)'.[11]

The concepts and practices of 'discipline' and 'training' link the modernisation of nursing and hospitals with both religious and military discourse. While Foucault takes the army as exemplary and paradigmatic of modern regimes of discipline, the connections with monastic/conventual discipline are quite apparent, and at times, are explicitly drawn.[12] In this respect Foucault and Weber converge. Both religion and militarism worked through notions of sacrifice and service, both demanded complicity with hierarchical obedience, and both aimed to create trained and disciplined bodies. It should not be surprising that invocation of the military was common in discourse on nursing, and accompanied a religious strand which ran through virtually all versions of nursing in this period. One woman told the International Congress on Charities, Correction and Philanthropy: 'The organization of a training school is and must be military... Absolute and unquestioning obedience must be the foundation of the nurse's work, and to this end complete subordination of the individual to the work as a whole is as necessary for her as for the soldier.'[13] Often, as Martha Vicinus points out, and as this 1907 textbook exemplifies, the military metaphor of management was joined by the analogy of disease as the enemy to be fought: 'No army can be successful in the field unless the general can rely on absolute obedience in all the different ranks, and no hospital can be successful in its campaign against disease unless the honorary staff can rely on absolute obedience all down the different ranks to the junior probationers.'[14]

Mary Poovey argues that in the context of mid nineteenth-century nursing, domestic and military narratives converged quite neatly, and without any necessary sense of contradiction. The domestic, she writes, 'always contained an aggressive component'.[15] Similarly, the discursive institutions of religion and the military, as they were taken up with respect to nursing, were often quite compatible. In a wider context, of course, religion, the military, and war had a longstanding connection. Notwithstanding the masculinity of this connection, there was some room within the discourse for women. More accurately, femininity worked in a symbiotic relationship with this masculinity. In war, in Jean Bethke Elshtain's words, men and women, masculinity and femininity, take on 'in cultural memory and narrative, the personas of Just Warriors and Beautiful Souls'.[16] Much of Anne Summers' study, for example, suggests the cultural ease with which 'British Women' could become 'Military Nurses'.[17] The space for the compatibility of religious, military and domestic discourses in nursing was enlarged immeasurably by the figure of Nightingale; by the almost equally weighted significance in the Nightingale narrative of religion (her 'divine calling', the Deaconesses' Institute at Kaiserwerth), the military (the Crimean War, her involvement with military hospitals, the Indian Army), and the domestic ('a housewifely woman').[18]

Nurses underwent a bodily training which drew on this conflation of military and religious discipline. It defined their way of acting, moving, gesturing, and standing in space, to quite minute levels. One text set out the following rules: *'Do not lean against beds... Get in the habit of walking quietly,* not on tip toe or any attempt at tip toe... *Get in the habit of looking cheerful, and speaking cheerfully'* and so on.[19] The author of 'Hospital Discipline and Ethics' wrote:

> Her voice, her laugh, her conversation, her walk, her touch, her habits of dress, the expression of her face; all tell their own story and bear on the question of her fitness and unfitness for the work she has undertaken... If all Nurses could be given a thorough

drilling in how to carry themselves, and how to acquire a graceful walk, it would be a distinct advantage to many Nurses who have unconsciously allowed themselves to become round-shouldered or awkward and ungraceful in their general movements.[20]

Discipline, according to Foucault, 'is a political anatomy of detail'.[21] And indeed the observance of detail was brought constantly to the attention of nurses. All and any aspect of their bodily presentation, cleanliness, behaviour, uniforms, were subject to observation and inspection, as were the minutely detailed diaries and accounts which recorded the management of their wards. However, the most powerful, and in a sense panoptical control, was imposed by the oft-repeated warning that there was always a smaller detail which could potentially be noticed by an observer who mattered: 'very few new probationers will notice the crooked quilt, the untidy locker, the crumbs by a child's cot or the orange peel thrown in the fender, that quite spoils the smart appearance of a ward'.[22]

The modes of command which Foucault explained, again in terms of the army, also apply to nurses in hospitals:

> All the activity of the disciplined individual must be punctuated and sustained by injunctions whose efficacity rests on brevity and clarity; the order does not need to be explained or formulated; it must trigger off the required behaviour and that is enough... it is a question not of understanding the injunction but of perceiving the signal and reacting to it immediately, according to a more or less artificial prearranged code.[23]

One 'prearranged code' which nurses needed to be trained to understand and enforce, was how to react, bodily, to the presence of superior officers, especially doctors: when and where they should stand in relation to the doctor, who should pass through doors first (in hospitals professional status overrode gendered etiquette), and when it was acceptable to be sitting in the presence of a doctor. In a 1912 lecture, for example, nurses were instructed that it was customary, 'when the medical officer in charge enters for the first time in the day, the dining or sitting-room in which the Nurses not in the wards may be sitting chatting, reading, or writing, for all of them to rise till the officer indicates if he is going to visit the wards at once, give a lecture, or is going to sit down and discuss with them matters of mutual interest'.[24] A textbook by the matron of the London Hospital details such regulation of bodies in the wards of hospitals. It also illustrates the panoptical surveillance of behaviour by 'those who are capable of judging', potentially, any observing individual:

> Never remain seated when visitors are passing through the ward... never think of sitting down or of remaining seated while you are speaking or being spoken to by *any* of the medical staff, whether a senior or junior member of it, including the dressers and students; it looks quite unbusiness-like and unprofessional. There is... no question of social equality or inequality involved in this; but if you forget it, those who are capable of judging will know at once that you are ignorant of ordinary details of hospital routine... On the other hand, take every legitimate opportunity of sitting down and of resting your feet, for you all have a great deal of standing.[25]

The time-regulation which Foucault identified as a crucial process by which bodies in modern institutions were controlled, was derivative of the strictly regulated conventual or monastic day.[26] Time was rigidly ordered around patterns of sleeping, eating, working, exercising and, in some cases, around set prayer-times. In hospitals, nurses functioned as the primary time-keepers of the continual twenty-four hour day. They embodied, as it were, time-keeping and time-passing, as it was they who most obviously marked the change – the 'shift' – from day to night and back again. The 'Time Table for Probationers' drawn up by the Nightingale Fund, stipulated not when the nurse was to work, but when she was to do everything else; eat, sleep, exercise, pray. Work went on perpetually around these activities.[27] In remarkable detail, rules and regulations encoded when particular individuals – matron, sister, nurse – could leave their ward or the institution, not when she was to be there. Punctuality was instilled with some force.[28]

It was an imperative of new hospital architecture that wards be arranged to facilitate a relentless observation of the patients, and that both the nurses and sisters who did this were themselves supervised and observed. In some designs, this literal super-vision was achieved by placing a sister's room or station between wards, with windows facing both ways. This maximised the number of patients and nurses under observation at any one time. That 'hierarchical observation' and '[coercion] by means of observation' which Foucault has identified in other contexts were literally built into reformed hospitals.[29] One doctor suggested:

Wards of a small size are generally objectionable, because unfavourable to discipline, inasmuch as a small number of patients in a ward, without the presence of a head nurse, more readily associate together for any breach of discipline... One sister is capable of superintending fifty patients if placed in one ward, or in two wards if her sitting room is placed between and communicates with them.[30]

Poovey quotes Nightingale's efforts to institutionalise such intense supervision, efforts which sometimes bordered on schemes for espionage. Nightingale favoured a particular version of pavilion-plan hospitals because 'the Military Superior, the Surgeon, the Matron, can at any instant pop in upon any ward of a Hospital... Remember that Ward-Masters, Orderlies, and Nurses require inspection as well as patients.'[31]

In all of these ways, the new nurse became perhaps the central signifier of modern discipline within institutional and public health. Given the centrality of such drilled nurses to the micro-politics and culture of institutions, they held great possibilities to symbolise and effect the disciplining aspect of modernisation of health. One nurse articulated this precisely, writing of 'an army of Nurses... spreading and growing and presenting for the world's use either a strong, trained and united body of workers or a weak, undisciplined, straggling, unservicable body'.[32]

RELIGION

On the one hand, nurses in institutions represented, indeed embodied, the process of modernisation: training, discipline, order, regulation, control, efficiency. On the other hand, I suggest that these new nurses equally represented and embodied very particular religious values which the process of modernisation was ostensibly displacing. In what was otherwise undeniably a secularising and rationalising domain of medicine, the modernisation of nursing was not about secularisation in any clear way. One major point which needs making, is the extent to which modern disciplinary practices were intertwined with the religious discourse supposedly being displaced.

In his analysis of nineteenth-century nursing as a classic example of a secularising vocation/profession, Turner emphasises its shift from a base in religion into a bureaucratic, state-controlled system, and in broad strokes, this is certainly the case. By the turn of the century, in certain respects, nursing was well on the way to becoming a secular and scientifically/rationally defined occupation for women. However, Turner fails to explain why, in the mid nineteenth century, when the regulation and 'healing' of bodies was supposedly losing its religious dimension, that is to say, precisely when medicine was coming to occupy the social space left by religion, the whole movement to create Anglican and Protestant nursing sisterhoods emerged. In the British Protestant context, this was not a continuity of any existing religious or medical or nursing practice, but was most decidedly an innovation. It was a deliberate and quite successful attempt to locate signifiers of Christianity and Christian 'healing' – nursing 'sisters' – within the context of modernising medicine. Indeed, this often occurred in the very institutions where medicine was becoming most modern, rational and scientific, most quickly – the large London voluntary hospitals.

In many ways, it is not the increasing secularity of nursing which is most striking, but rather the ongoing insistence on religiosity and questions of morality. Nurses in hospitals came to signify certain positions within religious debate; that is, within tensions between High Church and the various evangelical and Dissenting churches.[33] For example, the King's College Hospital had been established by men within the Tory High Church. Yet it was not they, but the Sisters of St John's House who in effect performed and represented High Church values in their daily work in the Hospital. As Perry Williams has noted, 'for King's College Hospital to be a Christian community, religiously correct nurses were necessary'.[34]

All of the sisterhoods aimed to care for sick bodies as a way to care for sick souls, an idea which was by no means limited to religious institutions, but which ran right through nineteenth-century secular reforms in England and elsewhere. The concept of charity, so central to nineteenth-century class society and to the reform of nursing, was also bound up with its meaning within Christian theology. Mary Stanley, daughter of the Bishop of Norwich, led a party of 'lady volunteers' and Catholic nursing sisters to the Crimea.[35] Driven by High Church beliefs and a strong sense that hospitals were a place for urgent female mission work, she wrote in her 1854 publication *Hospitals and Sisterhoods*: 'in this land of Christian privileges, we should surely not be satisfied with the fact that thousands are healed of their bodily infirmities, without inquiring how far their immortal souls have been tended at the same time.'[36] Decrying the current profligate and drunken hospital nurses, as she saw them, Stanley called for an extensive system of sisterhoods to administer the physical and spiritual nursing of the sick in hospitals. By contrast, Anna Jameson called for Protestant 'Sisters of Charity' rather more in domestic, than in religious terms. Yet her language clearly demonstrates the extent to which this domesticity rested on ideas of women's morality, spirituality and purity:

> As civilisation advances, as the social interests and occupations become more and more complicated, the family duties and influences diverge from the central home, – in a manner, radiate from it... The man becomes on a larger scale, father and brother, sustainer and defender; the woman becomes on a larger scale, mother and sister, nurse and help... woman... begins by being the nurse, the teacher, the cherisher of her home, through her greater tenderness and purer moral sentiments; then she uses these qualities and sympathies on a larger scale, to cherish and purify society.[37]

Although it became possible, even by the early 1870s, to speak of hospital nursing as a (secular) 'profession for ladies',[38] both religious and domestic ideals, as well as the complementarity of men's and women's work as articulated by Jameson, resonated in discourse on nursing throughout the century.

Part of the reason for this continuing mobilisation of religio-moral discourse in the domain of health, is the sense it made, and the possibilities it offered, for middle-class women. Femininity and masculinity fitted quite differently into concepts of 'work' or 'professionalisation', and consequently the processes of rationalisation and secularisation had very different implications for women and men. Put simply, middle-class women had a far greater investment in religion and morality as a way into, and a justification for, their paid employment or public position. It is important to note, also, how central such values were for the development of a feminist consciousness and activism in this period. For women, as argued in previous chapters, connection with some notion of morality often provided a way to authority. Much rested on, and much was invested in the perpetuation of this definition of their work. Notwithstanding the considerable contestation to religious discourse, 'secular' nurses often enthusiastically claimed and reproduced a culture steeped in religious meaning.

Connections between religious and secular nursing were consolidated through an inherited and constantly reproduced narrative about the heritage of nursing – about what nursing 'really' was. In endless retellings of stories about the origins of the work, early religious sisters and deaconesses were seen to represent authentic nursing values. Histories of nursing from around the turn of the century, which worked powerfully to constitute the identities of the women who read them, claimed and consolidated a genealogy which typically began with the work of religious women and healing women in early Christian times, moved through the depiction of (usually French) medieval nuns, German Protestant deaconesses in the early nineteenth century, and claimed Elizabeth Fry along the way.[39]

Secular nursing drew much of its meaning and its discursive practice from principles of chastity, obedience, and poverty. It was possible, though obviously problematic, for women working in hospitals to make some sense of their physical hardship, their strict institutional regulation, and their lack of remuneration, by thinking of themselves as types of nuns. The sacrifice of self, as will be seen, had a particular place within Protestant asceticism, which in turn, in Weberian analysis, was central to the development of secular professions. Yet in a quite different domain, self-sacrifice was one way, albeit an extraordinarily difficult way, for middle-class women to achieve status and recognition. Inasmuch as 'sacrifice' had a religious meaning, it had a gendered meaning as well. In the subjectivity of nurses, these meanings converged in a quite unique way, and did so with considerable intensity. London Hospital nurses were told that their work was 'the most Christ-like work a woman can undertake... sacrifice of self is involved in the very word "nurse"'.[40] Self-sacrifice as a way of constructing nursing was mobilised by a whole range of commentators and participants in hospital and nursing reform. While the very real demands for self-abnegation always had an oppressive side, the convergence of these religious and feminine meanings of self sacrifice also offered limited opportunities for nurses themselves to construct their work positively. Nuanced readings of the modernisation of nursing must acknowledge both the oppressive and enabling possibilities of a concept such as 'self-sacrifice' at work, it must analyse both modernising and gendered dynamics, or better still, the gendered dynamics within and propelling modernisation.

As Martha Vicinus writes: 'Nursing was to be transformed from the most menial of women's work to the most exalted through the commitment of pure and selfless women.'[41] Such representation drew on a tradition of female martyrdom which was in turn becoming increasingly inseparable from the figure of the nurse. Images of missionary-nurses going among diseased 'others' regardless of self, of 'heroic womanhood' in war, of middle- and upper-class women and Anglican sisters risking illness and death as they cared for the working-class sick in cholera epidemics, all fed into and produced this tradition. Of course, the positive construction of feminine religious sacrifice of self in the specific context of nursing rested overwhelmingly on the figure of Nightingale. Her representation as the embodiment of selfless heroic womanhood was by no means a slowly emerging phenomenon. The immediacy with which this image of Nightingale was produced and taken up during the years of the Crimean War, was only matched by its ongoing currency, well into the twentieth century. Nightingale certainly mobilised narratives of martyrdom around herself, as well as around other martyr-figures she created: Agnes Jones and Elizabeth Wardroper, for example.[42] By others, she was constructed as a kind of female Christ.[43] A contributor to the *Nursing Record* wrote that 'there are certain women who stand to us as beacon-lights on the hills of compassion, each one a Christ because animated by the Christ spirit'.[44] Nightingale was clearly aware of the power of the image of quasi-religious feminine self-sacrifice; a power she could not afford to dispense with entirely, either for herself, or for the 'cause'. At the same time, however, she wanted to distance herself and the institutions and reforms with which she was associated, from the work of religious women. She often expressed a profound disdain for nursing nuns.[45] Yet, as Mary Poovey notes, the uses to which 'Nightingale' was put, soon outstripped the control of the woman herself: '[H]er image had become far more powerful than anything she could say or write, and it had long escaped her control.'[46]

The hagiographical images of Nightingale which long sustained the connection between reformed institutional nursing and religion assumed many forms. Often, they were visual images: paintings, drawings, statues. Much of this was a self-authorised 'hagiography'. Nightingale sent statuettes of herself around the world. For example, one was sent to Lucy Osburn, the sometime Nightingale protegé, who wrote of the statuette:

> Mine I have had a bracket made for & placed over the chimney piece under the clock in my office. I assure you I am very proud of it up there [.] with the hand raised you look as if you were pointing the way to the Heavenly Kingdom. The Roman Catholics look upon it with great reverence [.] I suppose they think you have

been canonised. Do you know that both here & in London almost everyone thinks you to be R. Catholic?[47]

The letters written to Nightingale from the English women sent to the Sydney Infirmary suggest the common set of meanings between nursing and female religious communities. Osburn wrote, for example, of taking her 'vows' for six years of service to the Nightingale Fund.[48] Another informed Nightingale that 'one of her [Osburn's] first duties was to establish Family prayer for the Sisters, Nurses and Scrubbers, both morning and evening'. This comment suggests a familial domestic arrangement, but also hints at a more formalised community of religious women, in which work went on around set prayer times. The English sisters, initially at least, saw the lady superintendent's role as moral guide and exemplar. One wrote of Osburn with a reverence not simply reliant on differential class positions, but which resonated with religious adoration: 'it is such a blessing for us, that we have such a Lady to look up to as our head, we love her, and the more we know her the more endearing we find her. I am sure we cannot esteem her to [sic] highly.'[49] Another of the English women wrote of Osburn that her 'one aim...has been...to raise the tone of ideas & stimulate to pure motives and feelings', and commented that 'Miss Osburn is frequently styled "The Lady"'.[50] In such ways, while nursing was certainly becoming entwined with secular and scientific medicine, as well as with bureaucratic state structures, a firm connection with religion was sustained in the collective culture of nurses themselves.

The hard physical work demanded of women in hospitals at first appears to sit uneasily with efforts to feminise and moralise nursing. Yet, this also made some sense within a religious framework, of course to the great delight of hospital managers. Martha Vicinus attributes the success of the upper-middle-class leaders of British nursing partly to their use of an ideology of work discipline then gaining real currency, and writes that 'nurses...would benefit...from the purifying influence of hard work. The strong activist strain of Victorian religion found its natural outlet in images of lady nurses as Christian soldiers in the battle against disease and degradation.'[51] Salvation and redemption through hard work and physical pain drew from a Protestant ascetic tradition, as well as a Catholic concept of penance.[52] Common to both of these traditions was a general expectation that women work – in all senses – for others, not themselves. Religious women's lack of interest in, or need for 'earthly reward', exemplified paradigmatically for later nurses, sacrifice of self for a higher good. 'The work is hard,' wrote Honnor Morten in her 1888 *Sketches of Hospital Life*, 'but many desire nothing better than to have no time for thought of self, both brain and body being completely at the service of others.'[53] To a large extent, it was understood that nurses' physical work was to be rewarded not by their own monetary remuneration, as in normal capitalist understanding of 'labour', but by the return to health of the patient.

It was also the bodily repulsiveness of so many nursing tasks which constituted sacrifice. The authors of nursing textbooks commonly employed the nursing-as-Christian-practice discourse to explain away the most distasteful bodily duties which fell to nurses. Catherine Wood's text warned that 'no woman should offer herself as a nurse unless she is prepared to do real hard work, to practice incessant self-denial... to take her daily share in sights and occupations repugnant to every refined and sensitive mind'. And later, she advised: 'If you set before yourself the highest aim to be a Christian woman and a nurse, you will purify and ennoble all you touch.'[54] When writing of nurses' duties in laying out the dead, Eva Lückes asked young nurses to remember: ' "Whatsoever thing thou doest, to the least of mine and lowest, that thou doest unto Me"; and that thought, too, will help you to bear patiently and without complaint the most revolting part of your task. It *does* call for self-denial.' Like Woods, Lückes put the onus on individual nurses to ensure a religious (compliant and obedient) approach to the most difficult and distasteful aspects of their work:

> Nursing is work that should develop all that is best and highest and most womanly in you; and if you find this is not so, be sure there is something wrong in the spirit with which you are doing it. Remember that the profession which you have taken up... was of old intrusted [sic] to the holiest women, and they did not find themselves the worse for it.[55]

It was precisely the possibility of reconstructing bodily and polluting work within philanthropic and religious discourses which enabled the participation of gentlewomen and lady-nurses. In the same way that middle-class and upper-class women undertook 'rescue work' with prostitutes without either their status or their femininity necessarily coming under challenge, so it was possible for women to undertake distasteful bodily nursing work without necessarily risking moral pollution, even if the risk of physical pollution, of contracting disease, was high. Moreover, as with 'rescue work', if nursing could be conceptualised in a very specific way, the potentially polluting nature of the work could, in a sense, produce the purity and respectability of the women involved. The meaning of polluting work in general, and of bodily contact between female nurses and male patients in particular, could be transformed by shifting the issue well into a religious arena. This relied also, of course, on constructing the patients in a particular way. An 1876 article, 'Sick-Nurses' proclaimed that 'the patients entrusted to the care of these women are the chief treasures of a Christian nation, since they are both poor and sick'.[56] In many such accounts of reformed nursing, the class rather than the sex of male patients is emphasised. The image of this type of nurse was powerful indeed, and could render possible intimate contact between middle-class women and working-class men. Indeed, given the severe mid and late nineteenth-century constraints on middle-class women's employment, it was as yet *only* possible for gentlewomen to work within a charitable context and to nurse men who were 'both poor and sick'.

In a myriad of ways nurses were encouraged to subsume their personal identity within a collective identity, based on work for others, or work for some higher principle. In some institutions, for example, the 'sister' in charge of a ward lost her own name and assumed the name of that ward. The identity of these women, then, came to be quite subsumed within not only their work, but their workplace. This was intensified dramatically when sisters slept in apartments adjacent to their ward, and so had almost no private identity. The wearing of uniforms also negated individual identity, creating a discipline and regulation in uniformity: 'An orderly and a homogenous appearance has a distinct value of its own, in promoting the internal discipline and cohesion needful for concerted work.'[57] An early text on nursing identified precisely the two objectives of wearing uniforms: 'All the nurses must wear the same apparel, very like Roman Catholic Sisters, or Nuns, for the purpose partly of being known as nurses wherever they go, and partly for equality, in order to prevent jealousy among themselves.'[58] Uniforms produced the primary identity of the women who wore them as generic 'nurse', publicly distinguishable from other women, but more or less indistinguishable from other nurses. Of course uniforms also went some way to produce the confused identities of nurses, simultaneously signifying cultural links with religion, domestic service and the military. The sexual significance of nurses' uniforms is discussed below.

The possibility of constructing nurses' work as sacred, or religious, enabled the concept of 'obedience' to take on far more than a managerial or servant – mistress/master meaning. Some doctors, hospital administrators and nursing leaders used this possibility more or less strategically. One doctor appealed to both familial and religious duty when instructing on a 'sacred' obedience to the medical men: 'The nurse should – forgetting everything of self, regard every patient as a dearly beloved relative... and carry out the doctor's instructions to the letter, for it is her sacred duty to do so.'[59] A 1906 article – one of many describing and prescribing the 'Perfect Nurse' – wrote that complicity with medical directions and orders should be undertaken 'in the spirit of professional and Christian obedience'.[60] Efforts in the early twentieth century to create a set of ethical standards, a kind of code of practice for nursing, seized on both self-sacrifice and obedience as the highest measures of 'professionalism'. One such 'code of ethics' stated that a nurse should be willing to sacrifice her life, no less, when in charge of fever cases: 'In the case of an epidemic it is her duty to face the dangers and to continue her labours for the alleviation of the suffering even at the jeopardy of her own life.'[61]

In all of these ways – through appeal to Christian (and Christ-like) moral/physical 'healing', through the idea of self-sacrifice, of female martyrdom, of spiritual purity – the religious was sustained in what was otherwise a secularising domain of medicine. For some nurses, this offered real possibility of authority, responsibility and fulfilment. For others, it was an oppressive, constraining and exploitative demand which engendered considerable resistance. For those in superior positions in nursing and hospital hierarchies, appeal to the religious could clearly be very useful, and the most appalling exploitation went on, consciously and unconsciously, by invoking 'selflessness', 'hard work', and 'sacrifice'. For all these different experiences and deployments of religious discourse, it retained real currency and was sustained alongside other fields of meaning through which nursing was defined.

SEXUALITY

Sexuality, generally, was always problematical in modernising institutions and in the broad process of rationalisation. For order and rationality to prevail, unruly sexuality needed to be disciplined and controlled. Turner has written:

Disciplines of the Flesh 57

The irrational sexual impulses of the body constituted a special problem within the Christian opposition to the world... The rational control of these impulses through the institutions of celibacy and monogamy represents an important dimension of Weber's master concept of rationalization in which the emergence of labour discipline and asceticism in capitalism constituted a major historical turning point in the control of the body.[62]

But masculinity and femininity did not fit identically into such a process of rationalisation. In theorising asceticism, control of bodies, control of sexuality and so forth, it is often left unsaid that it matters which sexed body is at issue. In other words, a concept like ascetic control had quite different implications for female and male embodied subjects. On the one hand, there was a widespread nineteenth-century notion that men's sexuality was particularly difficult to control, that men were 'animalistic' in this sense. On the other hand, however, there is a tradition embedded in Western culture, of thinking of women *as* their bodies – defined and controlled by their organs, their sexuality, their corporeality. As many feminist scholars have shown, the dichotomy between mind and body is a deeply gendered one. Moira Gatens has written that 'the ideal conception of the rational is... articulated in direct opposition to qualities typical of the feminine'.[63] In women, the control of sexuality was the control of something always just submerged. As Poovey argues, anxiety over this instability was always the underside of nineteenth-century obsession with proper womanly behaviour.[64]

Women's particular construction in relation to their corporeal and sexual existence, as well as the particular place of female sexuality/temptation in Christianity, meant that femininity held a very different cultural relationship with the idea of 'flesh' or 'sin' or 'virginity' or 'celibacy', than did masculinity. Moreover, masculinity could not simultaneously signify sexuality *and* asexuality in the way that femininity could. In that 'woman' was so fundamentally constructed through the dichotomy of virgin and temptress, particular women could effectively represent the contradiction between the spirit and the flesh, so central to Christianity. The ambiguous symbolic status of the sexuality of nursing is a clear exemplification of this.

Like nuns, the new nurses were meant to be asexual beings. Yet it was this very purity and supposed innocence which came to be eroticised in certain contexts. While asexuality or religious chastity usually implied a lack of contact with other bodies as well as control of one's own, through which a refined spirituality might be attained, in the case of nurses and nursing nuns, asexuality was accompanied by intimate physical contact with male bodies. Both nurses and nuns broached the pure and the impure, the clean and the contaminated in a way which was charged with the possibility of sexual interpretation. If sex was constructed as dirty, nuns and nurses were constructed as pure. The sexualisation of nurses was about the transgressive violation of innocence, an eroticised taboo, a profane pollution of purity and sanctity. Once again, the pure and the polluted came together over the female body.

Such sexual ambiguity is perfectly exemplified in the meanings of the nurses' uniform. When late nineteenth and early twentieth-century nurses walked through city streets in what were known as their 'outdoor uniforms', they became 'public women' in several senses. They were troubled about being mistaken as prostitutes: 'They infest every night the public thoroughfares of London and other cities, bringing the deepest disgrace upon the uniform they wear'.[65] A correspondent to the *Nursing Record* demanded that 'something must be done, as, if the scandal continues, it may debar many from adopting a profession in which the recognised dress is no longer a protection but the reverse'.[66] The uniform was a sign which could be read different ways by different people: 'Rarely do we fail to recognise the masquerader ourselves, but unfortunately the public... cannot see with our eyes.'[67] For prostitutes, it was a 'disguise' in several senses. It gave them, as 'nurses', a reason to be walking alone on the streets, and the ability to be misrecognised. To potential male clients, however, the uniform was a sign of a particular kind of eroticisation, which rested not only on the recognition that the wearer was a prostitute, but also on the recognition of, and response to, an already eroticised image of the nurse. A nurses' uniform could be both a 'provocation and a disguise', warned Joseph Bell in his 1899 text *Notes on Surgery for Nurses*.[68] If nurses were linked to nuns as one aspect of their fragmented identities, then in another aspect they were linked to prostitutes.

There were several rationales for, and considerable debate over outdoor uniform, in which nurses' ('indoor') costume was seen as either a contaminated object or as something to be sanctified. Some saw the nurses' uniform as 'too sacred to wear anywhere but in the sick-room', seeing it, nun-like, as a symbol of purity.[69] For others, however, nurses' uniform represented dirtiness. They posed hygienic and public health reasons for outdoor uniforms, arguing that a nurse who wore her contaminated dress outside the sick-room was insanitary and disreputable, indeed, an 'unsafe person'.[70] Such contradictory argument again captures the uneasy symbolic positioning of nursing between purity and pollution, between sanctity and profanity, between a quasi-religious practice and a menial practice involving contact with contaminated bodies.

'Nursing', Poovey writes, 'both intimated and specifically sacrificed sexuality.'[71] The ease of transition from a nun's habit to a nurse's uniform to a prostitute's disguise, gives material expression to such a statement. This fluidity of costume held far deeper implications than the simple donning of particular garb. It both reflected and produced a multiply constituted identity, in which 'nurse' and 'woman' were closely, almost identically, aligned. Nursing could and did assume any construction of femininity or female sexuality which was available. In terms of the ease of transition from nun to nurse to prostitute and back again, for example, nursing reflected the way in which 'woman' always meant all of these things. It reflected the relative ease with which the lady can 'fall', the prostitute can be 'rescued' and turned into the innocent victim of rapacious male sexuality, the virgin can turn into the spinster-shrew.

The source of another contradiction in the domain of nursing, which caused some disquiet, was the nature of hospital/nursing environments as both intensely same-sex, in terms of institutionalised communities of women, and at the same time, intensely heterosexual, in terms of contact with male patients, and the dominance of the gendered/sexualised relationships with male doctors. While sexuality in nursing textbooks characteristically took the form of a very loud silence, every now and then an anxiety was articulated about single women's institutionalisation. A 1913 publication on the management of modern hospitals linked nurses' homes to other all-female communities, warning of the 'spread' of erotic tendencies, meaning either masturbation or lesbian sexuality: 'Training-schools are not exempt from those epidemics of morbidity that so often invade convents and young women's academies and girls' schools of every description. One morbid pupil, with erotic tendencies, can sometimes pollute a whole training-school.' Such illicit sexuality is classically pathologised, suggesting a submerged anxiety about all-female communities, as well as the ease with which even religiously-defined female communities could be seen to harbour a threatening, 'morbid' sexuality. It is significant also that this was followed by a discussion about the management of menstruating women.[72]

The troubling ambiguity of nurses' sexuality always represented an irrationality in institutions increasingly seeking out a modern, rational culture. This rendered nurses' discipline and training all the more crucial. Nurses embodied, and represented with some intensity, what Turner describes as a 'fundamental tension or contradiction within the Judaeo-Christian tradition which opposed the life of the spirit to the irrational dangers and temptations of the world... The contradiction between the life of the spirit and human embodiment is perhaps best signified by the notion of "the flesh".'[73] While the 'flesh' in Christian theology is what the pure spirit must escape, or overcome, nursing nuns and religiously defined nurses had to create and maintain their 'purity' at the same time as being immersed in a world of flesh and bodies. This further contradiction, written over the contradiction already inherent in their femininity, rendered 'nursing' a particularly intense site for ideas of 'fleshliness' and ascetic control to converge. It should come as no surprise, that there was such an explosion of texts on the 'training' of nurses, which variously mobilised, submerged, argued over or utilised nurses' embodiment of this deeply resonant tension between the body and the spirit. In this way also, the new nurse embodied discipline. But what is marked here is the extent to which this discipline was not a secular and rational discipline. It was a discipline resting on religiously resonant ideas about the female body, and female sexuality. This does not fit easily into models of modernisation which privilege the development of the secular and the scientific.

It is argued in Chapter 7 that doctors' pursuit of reductionist medical techniques and therapeutics – that is to say a scientific naturalism in which the sick body had 'simply' organic and physiological causes – was in part facilitated through nurses' assumption of older, non-reductionist ways of thinking about the body. In the context of the secularisation of culture and of medicine, nurses can be seen to play a similar facilitating role. It was in the discursive domain of nursing that ways of thinking about the body were perpetuated which allowed it to be seen as always more than a reified object. It was precisely in the field of nursing that some sort of morality, or metaphysical meaning, retained a purchase on medical care/control of bodies, albeit always tentative and provisional. The mutually defining nature of the worlds of nursing and doctoring is important in this scheme. Grounded in a very material connectedness, the femininity of nurses and the masculinity of doctors were produced specifically each in terms of the other. One of the most important mutual definitions of nursing and doctoring (although this was never entirely neat or uncontested) was the way in which one signified religiosity/morality so effectively, while the other signified secularity/rationality.

In the secularisation of hospitals and of medical care/control, nurses provided an ongoing link with religion. While in one sense this problematised the process of secularisation, in another it enabled and smoothed over the broad but massive cultural shift from religion to scientific, rational medicine. Nurses carried with them a latent religiosity and morality which could be mobilised or submerged with some ease. Its very flexibility was surely one reason for its continued presence. Qualities reminiscent of religion could easily become more benign feminine qualities, for example. Or, when it was appropriate for the author of a particular text, or the manager, matron or medical superintendent of a particular hospital, to distance medicine from religion, 'spiritual' nurses could become 'scientific' nurses or 'military' nurses at the flick of a few metaphors. Needless to say, the ambiguity of nurses' discursive position, located neither as fully scientific nor as fully religious, was the source of considerable confusion of identity for individual nurses, creating lives experienced through what was at times a very disabling contradiction. However, the extent to which this ambiguity was sustained and held currency, well into the twentieth century, suggests that in the unstable process of secularisation, rationalisation and modernisation, there was some investment in retaining a (benign because feminine) moral, spiritual or religious field of meaning. The new nurse functioned as a sort of cultural reservoir of religious–moral values in a modernising politics of health; a moral figure who nonetheless brought with her a modern rational efficiency. Nurses were modern versions of Christian 'healers', whose femininity, and overall subordination to medicine posed only usefulness, not threat, to a secularising culture.

4 Pathologising the Practitioner: Puerperal Fever in the 1860s

In the world of mid and late nineteenth-century medical practice, the issue of the bodily cleanliness of the practitioner came to be something of a measure of status and credibility. Increasingly, the ability to connect a particular type of practitioner to cleanliness was to gain authority for that practitioner, whether they be male doctors, female doctors, male or female nurses, accoucheurs or midwives. A sanitary discourse of purity and pollution, cleanliness and contamination came to be one axis around which debate on gender and medicine turned. Such dichotomous concepts lent very well to female bodies as they were imagined and represented and, as has been argued, the new nurses located themselves within this discourse. While there was a certain ease with which women could be thought of within this discourse of purity and pollution, comparative speculations were occassionally made of the bodies and the bodily cleanliness of male practitioners.

In this chapter I focus on one site and moment in which the male body became the object of scrutiny: not the male body of the patient, but of the practitioner himself. In debates on medical attendance at childbirth around the 1860s, male practitioners – accoucheurs or men-midwives, specialists in obstetrics, general practitioners – came to be problematised in several ways.[1] An explicit politics of sexual difference entered the medical domain, following heated debate on the use of chloroform in the 1850s.[2] The sudden activity around the question of female medical practice, including the education of women doctors and the training, expertise and responsibilities of midwives, inevitably rendered the masculinity of 'normal' practitioners visible and open to question. Part of this process, which both produced and was produced by discussion on female medical practice, was a questioning of the sexual morality of male attendants. The male practitioner – specifically the accoucheur – was not only 'sexed' in these debates, but was constructed as illegitimately sexual. In this process of 'sexing' the male practitioner, he came to be articulated as embodied in an unprecedented way.

A further way in which the embodiment of the male practitioner became discursively explicit in this period was within and through debates on puerperal fever. The battle between different types of practitioners to be 'clean' was played out intensely over the controversial statistics of maternal mortality. Male practitioners were accused not only of the sexual and moral 'contamination' of childbearing women, but of physically contaminating them with the contagion of the deadly fever. Here, in a radical reversal of the pathologised female body, the male body was scrutinised as, if not quite the source of pollution, then its conduit. There is an uncritically received story about all this that medical men refused to accept or even consider that they were the carriers of disease. Adrienne Rich, for example, writes of an 'outrage at the implication that the hands of the physician could be unclean; uncleanliness was the very charge the doctors had long been leveling at the midwives'.[3] Certainly there was some opposition to the empirical research which incriminated medical men. But careful reading of medical articles and textbooks reveals the significant extent to which the possibility that puerperal fever was a contagion carried by accoucheurs and other medical men was treated rigorously and seriously. Far from there being some sort of outright refusal of this possibility, the debate was complex and multi-faceted.

Through the puerperal fever debates, the practitioner came to be pathologised in a manner quite unique to the mid nineteenth century, but increasingly common by the turn-of-the-century. In thinking of themselves as possible carriers of contagion, and more importantly as asymptomatic carriers, medical men subjected themselves to experiments about their own bodily dirtiness or cleanliness, to close investigations of their bodily interactions with patients, to rituals of cleaning and to an obsession with hands which anticipated turn-of-the-century aseptic practices. While it remains the case that there was never a consensus about the aetiology of puerperal fever or the place of the medical man in its proliferation, the problem of puerperal fever certainly forced the medical gaze inwards.

GENDER AND PROFESSIONAL TENSIONS

In the middle decades of the nineteenth century, dominant sections of the English medical world were professionalising quickly. Yet the general medical scene was still disordered and in considerable flux.

In 1850, for example, there were 19 medical licensing bodies in Britain,[4] and despite the Medical Registration Act of 1858 which provided a semblance of internal order and alliance, even conventional medicine was still quite fragmented. Conventional medical education took place in a range of institutions and through a range of practices, from a traditional apprenticeship system, to lectures and demonstrations given by the Licentiate of Apothecaries, to the Universities. New knowledges and technologies meant that established hierarchies within the field were being disrupted. The higher status of physicians for example, while remaining broadly intact, was beginning to be challenged by the new successes of surgeons aided by chloroform. That professional cohesion which so many mainstream and particularly metropolitan medical practitioners sought was also challenged externally by a number of alternative practitioners, from homoeopaths to chemists and druggists to bonesetters and to midwives. As Ornella Moscucci has written, the agitation to reform regulatory procedures of practitioners in the early nineteenth century was begun by the apothecaries who wanted to exclude chemists and druggists from legitimate practice.[5] Regulation was always driven by the desire to exclude one group in order to establish the legitimacy and status of another, and the rest of the nineteenth century was to be marked by constant tensions and struggles between types of practitioners in the broad domain of medicine.

The field of midwifery practice was not the least of these struggles. It was not until the 1850s and 1860s that the College of Surgeons and College of Physicians opened their fellowship in any way to accoucheurs, and even then the moves were tentative indeed. Following the 1858 Medical Registration Act men-midwives moved to strengthen their tenuous position, and began to redefine their work with all the markers of professionalism: the Obstetrical Society of London was formed, an obstetric journal was published, and the moves to exclude any 'dubious' practitioners were intensified, first and foremost midwives.[6] In 1872, the Society set up its own scheme for examining and registering midwives, attempting firmly to control the knowledge and the practice allowed them.[7]

There was no clear picture and considerable anxiety over just how women and femininity were to fit into this professionalising medical world which was riddled with tensions between different interest groups. The various efforts by women in the 1860s to learn and practice conventional medicine, to radically restructure hospital management as new nurses or philanthropists, to continue independent and unregulated midwifery practice or to wrest the treatment of the diseases of women from medical men, loomed as immensely disruptive. The new specialist obstetricians were especially cautious, and women doctors like Elizabeth Garrett Anderson were specifically refused entry to the London Obstetrical Society.[8] It was nonetheless the case that the difficulties of female medical practice were not simply met with some sort of outright and unified opposition. Rather, there was a spectrum of ideas and enterprises around female medical practice. Many alternative male practitioners promoted female medical practice or independent midwifery vigorously as a way to maintain an assault on conventional medicine, thus preserving the unregulated, uncontrolled system which was in their interests. Some medical men closer to conventional medicine tentatively pursued the idea of women doctors. Others attempted to stabilise the system of conventional medicine by incorporating women into the modern, regulated scheme and dividing it hierarchically along gender lines: doctors on the one hand, nurses and midwives on the other. Another group suggested two spheres of practice complementing the two sexed bodies that get sick, with men treating the diseases of men, and women treating the diseases of women.

One group of men who were very interested in the possibilities of female medical practice formed the (all male) Female Medical Society in 1864. The public discussion this Society provoked demonstrates a real uncertainty over just what constituted female medical practice, as well as anxiety over how women were to fit into the emerging modern and professional medical world. This group of men – mainly doctors, but also churchmen, businessmen and a few members of parliament – all advocated female medical practice of some sort. Primarily through the efforts of Dr James Edmunds, temperance reformer, surgeon with the Royal Maternity Charity and later the London Temperance Hospital, a small school for the medical education of women was opened.[9] It was variously called the Ladies' Medical College, the Female Medical College, the Obstetrical College for Women and the Female College of Midwifery. This changing nomenclature, from 'Ladies' to 'Female' and from 'Medical' to 'Obstetrical' to 'Midwifery', encapsulates both the direction the College took from 1864 until its closure in 1873, and the general uncertainty over just what a female medical practitioner was or might be. Through its nine years of existence, this College did educate and certify many women, and its significance seems to have been eclipsed by historical interest in the London School of Medicine for Women, opened in 1874 through the efforts of Elizabeth Garrett Anderson, Sophia Jex-Blake and others. In 1867 the College announced that 50 women had enrolled, 10 widowed women, 15 married and 25 single women. These included midwives already in practice, hospital matrons, aspiring medical missionaries, a considerable number of widows, wives and daughters of medical men, and several women described as 'amateur students' who were to gain some medical education to assist their philanthropic work.[10] The debate which this venture sparked forced public pronouncements about women's work in medicine and midwifery from all the disparate groups concerned.

Initially, the Society had a broad agenda 'to facilitate the admission of Women into the Medical Profession, by establishing a Female Medical College in London, and to promote the due qualification and registration of Female Practitioners in Midwifery and Medicine'.[11] Also part of the plan, but never realised, was a hospital or dispensary in London staffed by female medical practitioners who would treat women and children.[12] By 1867 the Society was already backing away from this aim under pressure from physicians and surgeons at the large metropolitan institutions:

> The object of this society has not been to make 'lady doctors' who will undertake the diseases of men as well as of women... this Society has confined itself to giving instruction in midwifery and the diseases of women and children, and medicine as makes a sound and complete course of scientific instruction.[13]

And by 1872, just before the College folded, the emphasis was even more strongly on midwifery:

> The Treatment of the diseases of women is necessarily more or less associated with the practice of Midwifery. There are cases which call for all the resources of a full medical training... It is submitted that all professional Midwives should possess some knowledge of these diseases. Many women of delicate feeling would gladly welcome their assistance... some women suffer severe physical pain and distress of mind, and even permit their diseases to pass the curable stage before consulting a medical man... It has been decided that, in order to more precisely define the scope of the Society's teaching operations, the College shall henceforth be designated the Obstetrical College for Women, and it is hoped that it will work hand in hand with the Medical Profession.[14]

Edmunds and other lecturers at the school understood midwifery to involve more than attendance at childbirth, yet the phrase 'hand in hand with the Medical Profession' implied that the College now aimed to produce something other than doctors. While the College was heavily criticised, many wished to become involved precisely in order to contain women's medical practice. Dr H. W. Rumsey of the General Medical Council favoured the need for the licensing and training of midwives, but refused to support the Society fully, as it did not have in place the 'necessary limitations on the medical education, "qualification", and licence of women'.[15] The sudden interest in midwives was about establishing, in the words of the editor of *The Lancet*, 'a recognised system of subordination'.[16] And the sudden willingness to contemplate schemes and processes of regulation should be seen in part as a deflection of the threat of women becoming doctors.[17] For many such men it was far preferable for women to be trained through an institution under medical control, like that run by the Female Medical Society, than to allow the type of independent, private tuition with which Elizabeth Garrett had qualified for entry into medical ranks. Such control meant minimising the professional scope of subsequent female practitioners.

Criticism of the Female Medical Society also arose from some involved in the women's movement in the 1860s. John Stuart Mill wondered why its object was so limited, why a complete medical qualification was not attempted.[18] Elizabeth Garrett and her sister Millicent Garrett Fawcett thought that all efforts for women's medical education should be directed to already authorised and registered bodies and that precisely the same courses and examinations expected of men should be expected of women.[19] Florence Nightingale also opposed the medical emphasis of the college, the tendency as she saw it to produce lady doctors instead of midwives. The college also clashed with her own efforts to educate midwives at the King's College Hospital.[20]

A major premise on which the College was founded was the relative cleanliness and dirtiness of male and female practitioners. In the mid 1860s James Edmunds proposed that male doctors themselves were spreading disease through hospitals and homes. This, he argued, was the most important reason for the full development of a separate female medical and midwifery practice: as they did not partake in dissections of the dead, women were cleaner.[21] As has been seen the new, reformed nurses constructed their cleanliness, order and professionalism against the supposed filthiness and impurity of the old midwives. In no uncertain terms Edmunds loaded the polluting characteristics usually attributed to old midwives, the 'Gamps', onto the hospital surgeon. He depicted such men not only as the medium of contamination, but of death, as they moved from the scene of dissection to the scene of operation:

> There is also a fever which stalks through the surgical wards of our London hospitals... it assumes Protean shapes... 'Hospital gangrene', 'pyaemia', 'phlebitis', 'typhoid state'... It kills the great majority of all who die after cutting operations within the walls of our hospitals... Any day this week you may witness men who for hours have been handling half-putrid dissections, cast off their aprons and hasten to the operating theatre, there, with reeking hands, to manipulate the freshly incised surfaces of living, trusting, hoping fellow-creatures.[22]

These accusations provoked a sharp response involving law suits. Judging by subsequent debate in the medical journals, many doctors were more challenged by charges of contamination than by the initial issue of medical women itself.[23] While there was nothing particularly new about Edmunds' charges themselves – the question of medical men's role in spreading puerperal fever was quite widely discussed – he politicised the question by specifically comparing midwives and medical men. Moreover, he merged this issue with that of female medical practice and the medical education of women, provoking a heated response. Midwives also joined this debate, politicised by the recent attempts to regulate and control their practice. They questioned the damaging image of their own dirtiness by capitalising on statistics about relative rates of puerperal fever after male and female attendance. Isabel Thorne, graduate of Edmunds' Female Medical College, made much of the statistics and empirical studies which confirmed the infection of women by their accoucheurs, the numerous 'instances of puerperal fever propagated by a medical man'.[24]

I shall return to the question of puerperal fever later in this chapter, first to think about another way in which the accoucheur was constituted as 'dirty', that is as sexually and morally dirty. Midwives like Isabel Thorne challenged accoucheurs on moral grounds as well as on grounds of interventionist obstetric practice and the possibility of infection. The impropriety of having a medical man in the bedroom, often without the presence of the husband, was referred to again and again. 'Gentlemen! husbands! fathers!... [w]ill you not prefer to have the holy privacies of married life maintained in unsullied purity, undisturbed, inviolate?'[25] The 'invasion' of the accoucheur into this private architectural space stood for the imagined, but sometimes literal, penetration of the bodily space of the woman herself, either through gynaecological examination or through the rape of the woman. And certainly many cases of the rape of women by accoucheurs were brought to the criminal courts.[26] In such articulations, there was a clear reversal of the discourse which established midwives as dirty and morally corrupt. Instead the medical man was constituted as an embodied being, sexually, morally and physically dirty; there was a 'sexing' of the male practitioner. Nearly all those who pronounced against accoucheurs on the grounds of sexual indecency also argued that midwives be solely responsible for childbirth. James Edmunds saw one of the benefits of establishing women in the practice of midwifery, and of separating midwifery from surgery and medicine, as preventing 'those odious professional scandals [which] are now continually arising'.[27] If, as Christopher Lawrence has convincingly argued, the rising status of the medical profession had more to do with the cultivation of class-based gentlemanly attributes than scientifically-based technical expertise and knowledge, then it was imperative to distance the profession from such scandals.[28]

One major way in which the male medical practitioner was discredited then, was through constituting him as morally polluted, and setting this against the purity of women in childbirth. Accoucheurs were not only seen to pose a physical risk in terms of bringing puerperal infection, they brought a risk of moral contamination. In such arguments, the accoucheur was represented as a malevolent figure, practising (according to one opponent in an article 'Murder of the Innocents') 'butcheries of the Art... The accoucheur, if young, burns for experience; his new and glittering instruments, yet unfleshed, tempt him.'[29] They were seen as seducers who turned otherwise pure women from virtue to prostitution: 'man-midwifery leads to adultery and prostitution' declared one.[30] In a way which stands perfectly for the Victorian conflation of the physical, the moral and the sexual, the accoucheur was seen to perform each of these modes in the one act. That is, in the vaginal examination itself, the woman was seen to be morally, physically, and sexually contaminated. The hands of the accoucheur both violated the woman and, increasingly, were seen physically to introduce the puerperal fever. Consider this retelling of a rape case:

abruptly ordering the Husband and Protector out of his wife's chamber (a piece of medical presumption and insolence)...he frightened and disgusted his patient by his conduct and unnecessary interference with her, during a natural labour...The agony of her lacerated body caused her to bitterly complain of *his nails.* To her honour, be it said, she complained to her husband...and refused his further attendance, and was compelled to seek other aid in consequence of a violent attack of inflammation on the womb.[31]

If there was any part of the embodied male practitioner which was focused upon and problematised in this debate, it was his hands and fingers. While there is still in operation a powerful image of the historical obstetrician employing sinister instruments and the midwife using her benign and gentle hands, obstetrics in the mid-nineteenth century was most decidedly a manual practice:[32]

If the os uteri cannot be reached with the forefinger of the right hand, the two forefingers of the left hand should be introduced, as they can be passed higher than the forefinger of the right hand. Some distinguished obstetricians always use the second finger of the right hand, because of its length...In an examination in the early part of labour...the arm and wrist are rotated so that the palmar surface of the base of the finger comes almost in contact with the pubis, and the os uteri is explored with the radial and middle surface of the pulp of the finger.[33]

The particular dexterity required in obstetrics was a focus in many medical texts. Christopher Lawrence argues that there was a bid on the part of those in the highest ranks of medicine to reaffirm the practice as a skilled clinical art, an 'incommunicable knowledge', not the mere applied science implied in the use of the new technologies emerging from the continental tradition of medical research. Tools, he writes, like the thermometer, the sphygmomanometer, the ophthalmoscope, signified rather more the status of an artisan than a gentleman.[34] Those accoucheurs closest to the world of high ranking medical professionalism – those working in cultural proximity to the London voluntary hospitals – can be seen readily to embrace this rhetoric of 'the cultured practitioner of arcane skills'.[35] W. Tyler Smith, physician-accoucheur to St Mary's Hospital, London, wrote of obstetrics as an education of the sense of touch, such as is neither necessary nor acquired in any other department of practice. In tactile examinations of other parts of the body, sight and touch, or sight, touch, and hearing, are combined; but in the examinations of the accoucheur, touch is exercised without any assistance from the other senses. The obstetrician who has by long practice acquired the *tactus eruditus* in perfection may almost be said to have the end of his finger armed with an eye.[36]

The claiming of a *tactus eruditus* reads as a bid to construct obstetrics as a clinical art not a mere applied science, along the lines Lawrence suggests. Yet this was a risky strategy, for accoucheurs were thus trying to shift the inflection of the 'manual' nature of their work from the low status of 'labour' to the high status of 'clinical art'. Physicians, by contrast, were attempting to shore up their already highly valued clinical experience against the challenges of new continental technologies.

Unlike the physician, the hands of the accoucheurs were becoming highly questionable items. The touch of the accoucheur was unlike other forms of clinical touch. First, the accoucheur touched only women. Second, this was a 'blind' touch, to use Tyler Smith's terms, precisely because the accoucheur's hand was not on, but inside the body of the woman. Third, if in Mary Douglas's terms 'we should expect the orifices of the body to symbolise its specially vulnerable points',[37] this invasive touch took place at the most vulnerable orifice, at the most culturally loaded bodily boundary of all. In the 1860s, this touch came to be constructed as a contaminating one, morally and physically. In texts critical of the accoucheur, all the objections could be gathered together and invested in his hands: hands in places where they should not be; hands with particles of fever attached to them; hands which carried the fever from woman to woman. The hands of the accoucheur, indeed of any medical practitioner who attended women in addition to their other practice, came to be constructed as dirty in an unprecedented way.

Problematising the hands of the accoucheur, it seems to me, was an interesting way in which the male practitioner was foregrounded as an embodied practitioner and a sexed practitioner. Women's hands were not problematical in this way. Women were promoted, and promoted themselves, not only as cleaner and less likely to contaminate, but as smaller. James Edmunds wrote that 'the slender, delicate hand of a skilled woman possesses advantages in this vocation which the coarse,

Pathologising the Practitioner

bulky hand of man can never rival'.[38] And Isabel Thorne wrote: 'In consequence of the smaller size of woman's hand, operations which are often dangerous or impracticable when attempted by a man, are comparatively easy for a lady to perform'.[39] If surgeons and accoucheurs had 'reeking hands', female medical practitioners had 'slender hands', literally more fitted for the practice of obstetrics. In the following sections I want to investigate carefully this charge of 'reeking hands', the ways in which medical men dealt with the possibility that it was they who spread the dreaded puerperal fever from woman to woman.

PUERPERAL FEVER

Puerperal fever was one of those nineteenth-century diseases – like typhoid and cholera – which generated an enormous amount of public, professional and state concern. There were several reasons for this. The childbearing woman was the site over which heated professional struggles took place; a site where the relative purity of male and female practitioners was 'tested'. The exploding governmental discourse of statistics, morbidity and mortality rates, epidemiology, population surveillance and monitoring, as well as an ongoing medical fascination with the insides of the female body, all rendered puerperal fever a major medical and public issue. In the cultural imagination, puerperal fever figured prominently as a poignant site where birth and death occurred in the same event. Needless to say, the actuality of puerperal fever was tragic, and while its increase over the nineteenth century is the subject of some debate, the probability of contracting the fever and dying in a Victorian maternity hospital was high: childbirth was a dangerous event.

In the 1860s, the aetiology of puerperal fever was being theorised on a contagionist-anticontagionist axis or, (in the specific language of the decade) a contagionist-epidemic axis, with some sort of compromise between several causes usually being resolved upon. The aetiological concept of epidemic disease as it was used in this context included a range of factors, from an older humoral notion of balancing secretions and excretions to seasonal atmospheric changes, from the clustering together of bodies to the specifically miasmatic notion of foul and polluting air. Puerperal fever was sometimes equated with epidemics like typhus fever: 'The same causes that generate typhus fever – imperfect ventilation, foul air, an epidemic constitution in the atmosphere – will produce puerperal fever'.[40] Conceptualising puerperal fever as an epidemic disease brought into play those same spatial issues which rendered urban, institutional and domestic spaces problematic, as discussed in Chapter 1. Buildings, architectural plans, overcrowding and the accumulation of bodies, cubic feet of air and ventilation were the issues at hand. Puerperal fever, like so many other diseases, was often seen to be about how many and which type of bodies were brought together in what type of architectural space. The fever, *The Lancet* editorialised, 'is in direct proportion to the number of parturient females associated together in their parturient period, or breathing the same atmosphere... this disease finds its habitat in large lying-in hospitals'.[41] Effective sanitation and modern modes of hygiene were sometimes seen as the primary preventive measure: 'A lying-in hospital imperatively demands... that it be constructed on the best sanitary principles in regard to position, cubic space, free ventilation, good warming, &c.'[42] This was taken to the extent of radically questioning the ongoing support of large lying-in hospitals. In the common parlance of 'hospital healthiness' and 'hospitalism', the building itself was sometimes seen to 'cause' puerperal fever, as if it was alive with its own will and agency. One commentator argued that 'by simple suppression of lying-in hospitals this deadly manufacture would be arrested'.[43] Outbreaks of puerperal fever and the belief in its epidemic nature also threw into question the presence of maternity wards in general hospitals. After the new Florence Nightingale Ward in King's College Hospital was forced to close (due to the remarkable mortality rate of 1 in 15 in 1867),[44] it was generally agreed that the addition of such a ward to a general hospital was a mistake.[45] In the case of the King's College maternity ward, foul air from both the surgical ward and the postmortem theatre was seen as the major cause, but it is interesting to note, as discussed below, that contagion brought by the medical students was also much discussed.[46] Far from being seen as contradictory, explaining the fever through each of these factors was widespread. Margaret Pelling has suggested that this sort of aetiological model of 'complexity and sense of compromise' was far more common than models suggested by early historians, in which contagionism and anti-contagionism were seen as mutually exclusive.[47] The contemporary debate about causality was quite explicit in its admission both of ignorance and lack of consensus.

It was nonetheless the case that the possible infection of women by their doctors was widely and often boldly discussed. The concept of contagion implied the action of bodily contact, and was not necessarily linked to the Pasteurian notion of germs or microscopic living organisms. Thus as early as 1825 contagion was defined as 'the application of such miasm or corruption to the body by the medium of touch'.[48] The word and the distinct concept of 'contagion' was thought through with extraordinary care and caution in the puerperal fever debates, for much was at stake. A very particular set of issues, interests and ramifications distinguished it from, say, debate over the aetiology of typhoid fever or cholera. Proposing a theory that puerperal fever was contagious meant directly incriminating medical men in the death of those quintessential nineteenth-century innocents, the mother and the newborn baby. It was precisely because so much was at stake that contagionism received such full and careful attention in medical literature on puerperal fever.

It was rare for any aetiological discussion on puerperal fever not to deal with contagionism, yet few medical men were willing to argue that the disease was only and always contagious. This was not only because of the unsettling implications, but also because to do so would have been, epistemologically, very difficult. The idea that a disease might have a single and necessary cause was only just being hinted at in the mid nineteenth century. According to the medical historian K. Codell Carter, Ignaz Semmelweis was the first to propose that a single cause was required for the production of a specific disease, puerperal fever, and that this pre-empted germ theory proper, in which a particular germ became necessary, as opposed to sufficient, for a specific disease to arise.[49] This, Carter argues convincingly, was the real significance of Semmelweis's work.

Semmelweis specifically placed the obstetric work of male medical students against female midwives at the Vienna Lying-In Hospital, in which the ward attended by the former had a very high mortality rate, and the ward attended by the latter had a relatively low rate. His study suggested that medical students carried putrefying matter from dissected corpses to the women in childbirth, and that preventive measures should focus specifically on this problem: male students should avoid cadaveric matter and they should systematically cleanse their hands and disinfect them with chlorine. Semmelweis's research was also significant in that it was implicitly about gender, and was mobilised by some to affirm the growing story of dirty medical men and clean midwives. Contrary to the standard story that his ideas were rejected by his contemporaries, Semmelweis's work and the statistics from the Vienna Lying-In Hospital were referred to regularly in British discussions.[50] What was taken up in Semmelweis's work was not the aetiological principle of the single, necessary cause, but the empirical practicalities of hand-washing with chlorine.

For W. Tyler Smith, writing in 1856, it was distinctly unusual that doctors should wash their hands, a practice he noted, but did not specifically recommend: 'It is impossible to be too scrupulous, in a matter of such moment, and I have known some accoucheurs who, on entering a lying-in room, always wash their hands before making an examination.'[51] In a text for midwives published in the same year, there is no mention of washing hands at all, even though the neatness, cleanliness and orderliness required of midwives was emphasised over and over. Cleanliness in this (just) pre-Pasteurian discourse did not automatically invoke washing dirt away with water but meant, rather, order.[52] The increasing interest in washing which did appear from the 1860s in the domain of obstetrics anticipated the surgical obsession with hands and washing which characterised the turn-of-the-century aseptic mode of cleanliness.[53]

How then did the midwife figure in this debate on the aetiology of puerperal fever? As mentioned earlier, there was some recourse to established images of dirty nurses and midwives against which the accoucheur could define himself as clean. Robert Barnes for one wanted to open up the public possibility that it was midwives not medical men who were the 'poison-carriers'.[54] In a lecture which rejected Semmelweis's theory of primacy of cadaveric infection, he pointed an incriminating finger towards women. Midwives are articulated here as themselves abject bodies, dealing with excretions and filth:

> the personal habits of most nurses are not characterized by such scrupulous cleanliness as are those of men in general. They are apt to wear the same dress and under-linen for some time... They rarely indulge in the morning 'tubbing' or shower-bath. Shut up night and day in close attendance upon their patient, performing offices which expose them to direct and frequent contact with decomposing discharges and offensive excretions, ever inhaling the emanations from the sick bed, seldom getting out of doors, it is inevitable that their system and clothes get impregnated, saturated with matter which... must be in the highest degree likely to provoke in them active disease.

He suggests that nurses/midwives should be relieved every two hours to spend time in the open air, should wear white and should take a warm bath once a week.[55] Much of this is familiar from the discussions about reforming, ordering and cleansing general nurses discussed in Chapter 2. *The Lancet's* editorials on this issue were far more tempered, pointing only occasionally to cases where the fever was spread by women. It argued that 'it can serve no good or useful purpose to endeavour to make up for the inherently defective medical capacity of the weaker sex by urging the grave error that midwives are less apt than medical men to propagate puerperal fever'.[56] In general though, and despite medical men's control of the dominant discourse, there was no bald attempt to deflect attention away from themselves as infective agents by accusing midwives. Given the currency of the figure of dirty Sairey Gamp waiting to be mobilised in this debate, this is surprising. Perhaps any accusation of midwives was a confirmation of the theory of contagion: if puerperal fever was seen to be contagious through midwives, it could certainly be contagious through any third party, including the accoucheur. Midwives, on the other hand, could support a notion of the fever as contagious largely without incriminating themselves, because the empirical research had set up such a strong association between contagion and the specific practice of dissection. In some ways the contest was between the comparative polluting power of the cadaver with which the male practitioner came into contact and that feminine pollution the female midwife embodied.

While puerperal fever was clearly thought about on the contagionist-epidemic axis, there was another mode of classifying puerperal fever: the classification of the disease as endogenous or exogenous, as arising from either inside the woman's body, or from outside. The well-established discourse of woman-as-disease, of disease arising *de novo* from the woman's body, is clearly evident in this debate on puerperal fever, and functioned as one way in which doctors denied the possibility of their own contamination. Dr Snow Beck, for one, wrote of 'septicaemic or pyaemic puerperal fever arising in the patient's own system'.[57] Elsewhere the 'self-poisoning' of women was insisted upon.[58] From the mid 1860s Robert Barnes of St Thomas's Hospital and the Royal Maternity Charity developed his longstanding categorisation of puerperal fever into internal and external causes.[59] In an 1882 edition of his text *Antiseptic Midwifery*, the classifications of 'endosepsis', 'autosepsis' and 'exosepsis' were still used, indicating respectively 'self-empoisonment' from a lack of balance or even from emotional disturbances, self-empoisonment from the absorption of matter left inside the woman after birth, and empoisonment from matter introduced into the woman from the outside.[60]

The boundary of inside/outside the woman's body was the basis of one type of aetiological classification, a classification peculiar to puerperal fever. But the boundary between inside and outside was not a straightforward or necessarily clear one. While puerperal women themselves might well have identified the vaginal orifice as the boundary between inside and outside, this was not the way the female body was constructed in medical discourse. The hand of the accoucheur could reach into the vagina and indeed into the uterus, and in some cases this was claimed as their territory, as still 'outside'. The real inside of the female body was often identified as that beyond the barrier of the uterine wall. In lectures by Robert Barnes the female body was represented as folding in on itself, creating an 'inner surface'.[61] In another articulation of this idea, this 'internal surface' was itself seen as unsealed, permeable or absorbing: 'a vast solution of continuity...[like] the gaping...open-mouthed vessels of an amputated limb'.[62] Nothing about the female body was sealed or solid and even this 'inner surface' was an 'absorbing surface'. Such a conceptualisation was made possible by thinking of the vagina and uterus not as orifices which mark the boundary between outside and inside, but as continuations of the external, legitimating the purposeful place of the medical hand. On another reading, the medical hand is consumed by the vagina dentata. It is precisely this formulation of the female body as permeable, leaking and absorbing which feminist theorists have been interested in theorising and criticising.[63]

PATHOLOGISING THE PRACTITIONER

In his Foucauldian writing on public health, David Armstrong has recently schematised the shifting ways in which lines of hygiene have been regulated over the past two centuries. Well established by the beginning of the nineteenth century was the notion of quarantine in which individual bodies were separated, isolated, subject to rituals of exclusion, the *cordon sanitaire*. Mid-nineteenth-century sanitary science introduced a new 'zone of hygiene' in which the space between inside and outside each human body was regulated and monitored; that zone which separated anatomical space from geographic space. Early twentieth-century public health produced a dominance of the concept of personal hygiene which monitored exchanges between people, between anatomical spaces. And what Armstrong calls the 'new public health' of the last two decades has 'discovered danger everywhere'. The shift has been from the wielding of sovereign power and physical control to a disciplinary control in the Foucauldian sense, in which individuals were subject to the subtleties of panoptical surveillance and self-surveillance. An earlier system of 'binary division' in which a literal geographic line was drawn between clean and dirty, gave way to a 'multiple partitioning' in which surveillance and regulation was dispersed widely and deeply: a public health has emerged in which any individual is always potentially ill, unclean or diseased.[64]

The management of puerperal fever in the 1860s is interesting in this scheme because several of these strategies and mechanisms of regulation were at work. Versions of *cordon sanitaires* physically isolated the 'clean' puerperal women from the unclean. But concern for the interaction between the bodily space of the puerperal woman and the architectural/ environmental space around her also implies the functioning of Armstrong's newer zones of hygiene. Further, the management of puerperal fever anticipated twentieth-century schemes of public health, in which exchange and contact between two always potentially contaminated bodies came to be closely monitored. But what was quite unique in the ways in which puerperal fever played out in these public health discourses was the problematisation of the body of the accoucheur himself.

Old methods of quarantining were always about spatially separating polluted and pure zones. The physical barriers and the policing of borders of all kinds, was always two-sided. Both those 'inside' and 'outside' were policed: the quarantining strategy could primarily be about 'protecting' the pure or about 'excluding' the polluted. Well-established strategies of physical isolation and exclusion were still employed as a response to the puerperal fever problem in the 1860s. The uninfected childbearing woman was protected by physically separating maternity wards from general hospitals and by building isolated cottage hospitals in which no other condition was treated. The ongoing endorsement of birth at home was also a way of keeping the pure and polluted away from one another. Bodies other than the childbearing woman were also policed, and particular persons were restricted entry into those special wards or hospitals, those 'pure zones'. What is significant about puerperal fever is that medical men and accoucheurs themselves came to be spoken of as the suspect bodies to be regulated and monitored. James Edmunds' idea, for example, was to firmly and physically separate the clean practice of midwifery from the dirty practices of other branches of medicine, and especially surgery.[65] That is, he wanted to quarantine the dirty male medical body, as he saw it, from the precariously clean puerperal body; he wanted to keep childbirth and confinement as a sort of uncontaminated zone.

In the dynamics of separating polluted and pure bodies in the maternity ward, of keeping contaminants away from uncontaminated sites, male medical students in particular functioned as the dangerous body, whose movements were to be controlled. In the King's College Hospital episode, for example, it was the medical students not the student midwives who were singled out for scrutiny. From the beginning the entrance of medical students in the Nightingale Ward was forbidden, in theory at least.[66] This focus on students is interesting. If the new nurses and midwives were coming to represent order and discipline in institutions, medical students represented a disorder, albeit of a different kind from that of the 'old' nurse. Medical students were reputedly, if not actually, riotous and ill-behaved. As one commentator put it, 'the very name of medical students is synonymous with all that is unseemly and disreputable'.[67] Moreover, medical education was still quite haphazard and scattered in the sense that a student gained different instruction from different sources: an anatomy teacher here, a surgeon there; taking one set of lectures in a private anatomy school, another at the University, a third as an apprentice, a fourth in a lying-in charity hospital. They were, in a sense, floating bodies, under no one's specific or overall control. This arbitrary and uncontrolled crossing of clean and unclean borders was contrary to the quarantining mechanism at work in the management of puerperal fever. In this sense, then, it was medical students not student midwives who were problematic, and whose movements between zones deemed hygienic and unhygienic needed policing. There was also a sense in which medical students were easy to blame. In the same way that the sexual abuse and maltreatment of working-class women in hospitals was constructed as an issue specifically about medical students, as if the medical hierarchy was innocent,[68] so in the context of institutional puerperal fever an emphasis on medical students functioned to deflect attention away from the obstetric specialists themselves.

In addition to the mobilisation of quarantining strategies, the management of puerperal fever around the 1860s was defined by a discourse of hygiene which monitored zones between anatomical and geographical space. Concerns about ventilation and overcrowding were products of classic public health in its sanitary science phase.

But close examination of the debate reveals the extent to which Armstrong's third phase – the supposedly early twentieth-century 'personal hygiene' phase in which exchange between bodies was cause for concern – was being worked out from the middle of the nineteenth century onwards. As well as focusing on the exchange between a single anatomical space – one person – and the geographic space in which this person was located, medical discourse on puerperal fever problematised two anatomical spaces – two people – between which polluted matter could pass. Once again, in the context of puerperal fever, this problematic exchange was not simply about sick bodies, but was primarily about the interface between the patient and the practitioner. W. Tyler Smith wrote:

> I believe that the blood of the accoucheur may take up a dose of puerperal poison without manifesting any special results in his own system, and that he may communicate it through the medium of the lungs to his patient. In the case of a poison so subtle, the air we breathe unites the circulations of the accoucheur and patient, and renders them, as it were, one.

He suggested inhaling 'the diluted fumes of chlorine several times a day'.[69]

Mid-nineteenth-century responses to puerperal fever were such that the corporeality of the practitioner became a site of the medical gaze. The gaze was turned inward and the male body itself came to be seen as pathological, the object of experimentation, scientific speculation, even diagnosis, and increasingly the object of preventive measures. The idea that the practitioner might be an asymptomatic carrier intensified the necessity of constant and vigilant monitoring of his personal hygiene and indeed required the implementation of his own therapeutic regime. Self-surveillance was impressed upon doctors. They gradually became part of the 'problem' of public health, part of the monitoring, investigating and gathering of statistics which was as administrative and governmental as much as it was scientific.

One manifestation of this in-turned medical gaze was the curious new phenomenon of the case history of the doctor, which began to appear alongside the case history of the patient. Fleetwood Churchill's 1866 text, for example, detailed the 'cases' of particular medical men whose practices were plagued by puerperal fever. The intricate detail of the normal medical case history became a documentation of the medical man's personal habits, his spatial movements, dress, time spent between patients. Some sort of diagnosis about the cause of disease – contagious or epidemic – could be deduced. The statistics and stories gathered offered minute information about the doctor and his practice, and only incidental information about the sick or dead women.[70] One accoucheur was reported as taking the following precautions: 'he caused his head to be close shaved; he entered a warm bath and washed himself clean; he procured a *new wig, new clothes, new hat, new gloves and new boots.* He did not *touch* anything he had worn, and took the precaution to leave his pencil at home, and his watch.'[71] After his detailed examination of such 'case histories', Churchill recommended the following preventive measures, all of which asked the accoucheur himself to attend to his own body, to his own personal hygiene:

> He should... change every portion of his dress, and wash his hands in a solution of chloride of lime, as well as in soap and water. Dr Semmelweis's suggestion of paring the nails close, is worth adopting... At the termination of each visit to a patient in puerperal fever... the hands should be carefully washed with soap and water before leaving the room, and his clothes repeatedly changed and well-aired.[72]

Those medical men convinced of the contagiousness of the disease were still unsure of the nature or the process of the contagion and precisely how their own bodies might be implicated. Was it something that did cling to hands or was it, in one doctor's words a 'birdlime', a 'halo' a 'nebula', a 'poisoned cloud'?[73] Such ideas increasingly implied the possible contamination of the practitioner's entire body, inside and out. The idea of fumigating the whole body of the doctor in a sort of disinfecting sauna arose: 'If the obstetrician has been attending a case of puerperal septicaemia, or indeed septicaemia occurring in anyone else, he should place himself in a small room or closet, and place some scales of iodine over a spirit-lamp, and allow himself to be surrounded by its fumes.'[74] Later, when it was known that germs were invisible and could be anywhere, aseptic techniques required a similar process of total sterilisation of the surgeon's body. But the epistemological grounding of these two practices was quite different. In the earlier discussion, ideas about disinfecting the whole body of the practitioner arose not because medical men *knew* that germs were everywhere, but because they *thought* the contagion might be anywhere; in other words minute and detailed rituals of cleansing emerged precisely because of a confessed ignorance about where or what the contagion of puerperal fever was.

The debates on puerperal fever introduced a whole range of new practices, problems, and ways of thinking about disease into the British medical domain. The puerperal fever debates suggested a shift towards that 'single factor reductionism' and 'necessary and specific causes' which emerged in the middle of the century and which characterised late nineteenth-century germ theory and twentieth-century aetiologies.[75] Antiseptic washings were discussed and often practiced, and although washing walls, floors and even patients with the likes of chlorine, carbolic acid and lime was not unusual around the middle of the century, the imperative for the practitioner to wash certainly was. The turn to self-speculation and self-surveillance in the middle of the century introduced a corporeal awareness on the part of the obstetric practitioner which became fully characteristic of other fields of medicine and surgery only around the turn-of-the-century. Notwithstanding the intermittent resistance to the idea, accoucheurs and obstetricians were really the first group of conventional practitioners to think of themselves as dirty, as potentially contaminating. They began to speak of the need for 'self-purification' in a discourse which had previously been contained to the problematic bodies of female nurses, and which was to become the 'aseptic cleanliness' of that zone of absolute purity, the operating theatre. The whole idea of there being an asymptomatic carrier of a disease implied the need for a model of disease as unseen but ubiquitous, a model in which 'disease is everywhere'.

5 Feminising Medicine: The Gendered Politics of Health

At the 1867 annual meeting of the Female Medical Society the American doctor Mary Walker was invited to speak.[1] In the face of much controversy, Walker was touring England, lecturing as much on dress reform and bloomerism as on women's medical practice. This tour came on top of shocking reports about Dr James Barry, the eminent British army surgeon and lifelong cross-dresser, who was discovered to be a woman on her death.[2] Such incidents in the 1860s added fuel to shrill pronouncements about the unsexing of women which medical education would entail, descriptions of them, for example, as curiosities which the public would wonder at, 'just as it wonders at dancing dogs, fat boys, and bearded ladies'.[3] In the feminist *Victoria Magazine*, editor Emily Faithfull worked hard to undo what she saw as the damage of Mary Walker's lecture tour to the cause of women's medical practice. The respectable figures of Elizabeth Garrett and Elizabeth Blackwell, as symbols of the new movement, were hurriedly re-inscribed with feminine meanings. Faithfull described Mary Walker as

> such an unfortunate contrast to ladies like Miss Blackwell and Miss Garrett, whose scientific knowledge and medical skill are only equalled by their modesty and quietness of demeanour... a woman may be a doctor, and yet retain the modesty, purity, and grace, which are her special characteristics.[4]

Middle-class women who wanted to practice medicine had to find ways of distinguishing themselves from male doctors on the one hand, and from working-class female practitioners on the other. To identify with regular male doctors, to be scientific and professional, was an effective way of marking themselves as different from working-class women who practiced abortion as much as midwifery. Yet to be too 'scientific' and 'professional' was to be too masculine, and to risk accusations of being unsexed and unnatural, accusations regularly directed to Mary Walker and of course to James Barry.

One effective way to begin to negotiate all these contradictory positions – to be scientific, feminine, professional and middle-class, to be different from regular male practitioners and irregular female practitioners – was to write women doctors into the same discourses which were then being mobilised to support middle-class women's work as nurses. Like nursing, doctoring for women was often depicted in relation to a set of ideas and practices around morality, domesticity, philanthropy, vocation, charity, spirituality and self-sacrifice rather than ideas around scientific knowledge, career or professionalism. Women who wished to work in the medical domain inevitably encountered these latter masculine discourses, and there was certainly some questioning of the supposedly given masculinity of biomedical knowledge and practice, as well as a growing feminist appropriation of notions of 'career', 'independence' and 'professionalism' for women. However, in the pursuit of women's medical practice in Britain, most effort was not put into inverting the gender order, but to extending the cultural boundaries of femininity itself. Countering the images of women doctors as unsexed and unnatural 'curiosities', Elizabeth Garrett strategically pursued all the markers of femininity, writing to Emily Davies: 'I feel confident now that one is helped rather than hindered by being as much like a lady as lies in one's power.'[5] While opponents of women's medical education chose to imagine and depict women doctors in the most masculine contexts – typically in operating theatres or dissecting corpses – advocates of women doctors most often depicted them in contexts where medical work could be seen as a vehicle for women's benevolence, moral goodness and influence: mission work, philanthropic work and the domestic and moral imperatives of sanitary reform. While the histories of nursing and of women doctors are usually analysed quite independently I explore here the extent to which, and the ways in which debate around these two issues – women as nurses, women as doctors – referred to and influenced one another. They were largely part of the same debate which drew much of its meaning from the gender/class configuration of the discourse on public health and sanitary reform discussed in Chapter 1. In the first section of this chapter I draw out the marked material connections between nursing and doctoring for women as well as the ways in which they were discursively linked. In the second section I examine the articulation of an oppositional 'feminine' medicine and its connections with Victorian feminism.

NURSES AND WOMEN DOCTORS

In 1860 the Nightingale Fund embarked on its enterprise of training nurses at St Thomas's Hospital, and in the same year Elizabeth Garrett, the first registered female doctor educated in Britain, embarked on her medical career. She did so by becoming a nurse at Middlesex Hospital, London. Various medical friends and advisers suggested that six months as a nurse on a surgical ward in a large London hospital would 'test' Garrett's desire and aptitude for medical work. This was, however, the only way for a woman to gain clinical hospital experience in Britain. Garrett detailed her daily routine to Emily Davies as well as what she learned in her initial months at the Hospital, not from the doctors so much as the Sisters with whom she worked: 'I begin at once to prepare for the dressings by spreading the different ointments, preparing the lint, lotion, poultices, bandages &c. While I am doing this at a side table the sister is going round & examining all the wounds &c. The simpler cases she leaves entirely to me very often, but the more difficult ones, such as cancer, she dresses herself while I look on.'[6] Garrett planned to spend three months at Middlesex, followed by three months at St Thomas's with the Nightingale Fund. She soon found that her position needed clarifying and some acknowledgement given to her desire to be not a nurse, but a medical student:

> [I]t will not do to go on long in the false position I now occupy at the hospital. I am nominally a nurse, but without any duties, no regular scene of action even. I may go into any ward & learn anything & do anything. As each ward has its sister & sufficient staff of nurses it is difficult to get much actual nursing work... Dr Willis treats me as a pupil & the house surgeons do the same... It appears to me that I should not go on receiving instruction as a pupil under the guise of a nurse.[7]

In pursuing this nursing experience, Garrett was also heeding the advice of Elizabeth Blackwell. In 1860 Blackwell published practical suggestions for women aiming to study medicine. A four year plan was advised: one year studying medical books with a tutor, followed by six months as a hospital nurse, six months in a laboratory, eighteen months gaining a college education in America and six months gaining midwifery experience at La Maternité in Paris, a maternity hospital which also educated and trained midwives.[8] In working as a hospital nurse, Blackwell advised women to set aside any pride or assumption of superiority, and clearly felt that much would be gained:

> [L]ater, it will belong to your duty as physician, to superintend nurses and carefully attend to the hygienic and other arrangements of the sick room... As also the prevention of disease and care of health is half the physician's work, all experience which bears upon these subjects will be of great use.[9]

Despite Florence Nightingale's opposition to the idea of women doctors, discussed below, she nonetheless invited Elizabeth Blackwell to direct the Nightingale Training School at St Thomas's Hospital. Blackwell's understanding of medical therapeutics as essentially a preventive and moral sphere of work complemented Nightingale's ideas about women's role as nurses. However, the offer was refused, Blackwell confiding to Barbara Bodichon that she simply could not abandon her 'medical plan', and also expressing a caution about the implications of working with Nightingale: 'I had the distinct perception that she would work me to death if I had accepted her offer.'[10]

Both Garrett and Blackwell were very determined to be recognised as doctors and not simply midwives. When Garrett opened the St Mary's Dispensary for poor women in London in 1866, she and the committee decided to leave attendance at childbirth, the 'easier departments of practice', to midwives, and so not 'diminish her usefulness as a physician for women'.[11] However other women, equally convinced of their desire to pursue work in the domain of health, hygiene and illness, were less sure of the distinction between being a nurse, a midwife and a female doctor. For example, the status of the women passing through the College run by the Female Medical Society was quite ambiguous. Similarly the career, aspirations and capacities of Lucy Osburn were marked by ambiguity. Before Osburn entered into a contract with the Nightingale Fund and accepted the position as lady superintendent at the Sydney Infirmary, she was negotiating an appointment with the Delhi Medical Mission as a female medical practitioner. In Delhi she was to treat sick Indian women who refused to see male doctors. She was to establish an institution for the training of midwives, and was to gain access thus to Indian women otherwise quite beyond the reach of English and Christian influence. As the director of the Delhi Medical Mission wrote: '[T]he Native ladies refuse to be seen by a doctor till too late, & thus great numbers die or pass miserable lives of pain & distress...

Miss Osburn would thus also have means of both directly & indirectly getting Christianity before her patients who would otherwise be utterly inaccessible to religious instruction.'[12] For some years previously she had been involved in clinical work in hospitals in other countries, and had gained considerable medical knowledge. At the Jews Hospital in Jerusalem, the Nightingale Fund were told that 'she attended the operations taking notes of the cases & assisting in the dressing & nursing the patients (almost daily) for three or four years'.[13] It was the concern of those grooming Lucy Osburn for her position in New South Wales to ensure that she would contain her medical knowledge and limit her interference in the medical domain. Henry Bonham Carter wrote to Nightingale: 'I hope you have impressed... upon Miss Osburn the importance of avoiding any semblance of an approach to medical practice even by suggestions to the medical men on particular cases. She has an army of medical books with her & I feel there is a strong temptation to many women to follow Miss Garrett.'[14]

Women's practice in colonial contexts often allowed them far more scope than did practice in Britain, and it was in India particularly that the ambiguous 'female medical practitioner', simultaneously doctor, nurse and midwife, was in demand. Problematically, exotic contexts and the treatment of non-English bodies gave women a certain freedom to undertake some medical activities which they simply would not have cultural or institutional permission to pursue in England. So, Osburn wrote to Nightingale of her planned work in Delhi that 'in as far as it was medical work in India I did not object'.[15] Medical mission work also served to legitimate and render viable and respectable the woman doctor. To premise the need for medical women on the figure of the Indian woman caught in the dark and closed physical and spiritual space of the zenana was to appropriate powerful cultural codes of philanthropy and colonialism, both of which had significant room for the play of women's moral authority. The movement for women's medical education came to rely heavily on the imperative of 'helpless' and 'suffering' Indian women's health in order to expand the scope of British women's practice, to increase the standards of their medical education and qualification, to increase their numbers, and to open women's hospitals and medical schools. Nightingale herself, when she eventually came to support the medical education of women, wrote: 'You want efficient Women Doctors for India most of all whose native women are now our sisters, our charge (there are at least 40 missions who will *only* have *Women* Doctors).'[16] The prominent doctor Mary Scharlieb narrated her life through precisely these discourses, taking up midwifery in Madras and later graduating from the Madras Medical College, in order to relieve 'the great and unnecessary suffering of the women of the country'.[17]

Stories about women's attempts to gain entry to university medical schools dominated nineteenth-century accounts of the 'medical women's movement'. Indeed contemporary historians insistently reproduce such accounts of women's entry into university medical education. However an alternative story circulated in the late nineteenth century, which constructed care of the sick as essentially and naturally feminine. In this account, male and not female doctors were seen as the aberration; women had long been doctors. Emily Faithfull, for example, thought that the question should be 'not that women wish to practise medicine, but that they ever discontinued a work so essentially womanly'.[18] In his pamphlet *Female Physicians*, published in the *English Woman's Journal* and later by the Female Medical Society, the American doctor Samuel Gregory wrote: 'Women always have been and always will be physicians. Their sympathy with suffering, their quickness of perception, and their aptitude for the duties of the sick room, render them peculiarly adapted for the ministrations of the healing art.'[19] In the same way that many women involved in the wider domain of sanitary reform insisted upon a gendered division of labour, divided at the boundary of the home, advocates of women doctors also made a concerted effort to render medicine a domestic practice. They attempted to recast medicine not as a scientific endeavour, but as an extension of a mother/wife's nursing duties and skills, claiming it as part of women's special sphere. This depiction of women's role in health in an 1865 issue of *Alexandra Magazine* shows the extent to which women's authority could be asserted if medicine could be contextualised domestically. Caring for the sick, the author wrote:

> is beyond their [men's] control: wives manage it; no matter how men assume to dictate, the matter is outside their sovereignty... Their health, diet, digestion... are at women's mercy, and in sickness their very reason is surrendered to her discretion; her hour of triumph and men's submission to medicine are synonymous; then comes the moment when she rules him, and boldly shows she rules, through dose, and lotion, bath, and blister.[20]

The image of women as mothers/nurses/nurturers was explicitly drawn on in the discussion of female medical practice. Given the prominent discussion on nursing, as well as the publicity around Nightingale and the Crimea, the cultural linking of women with the sick and with sick-rooms gained real currency in the 1860s. And notwithstanding Nightingale's objections to women's medical education, her image was invoked many times to legitimise the idea of women doctors. Elizabeth Garrett wrote to Emily Davies that when she announced to her father her intention of studying medicine, he responded that 'the whole idea was so disgusting that he would not entertain it for a moment. I asked what there was to make doctoring more *disgusting* than nursing which women were always doing, & which ladies had done publicly in the Crimea. He could not tell me.'[21] In a response to the argument that women were physically unfit for medical work, Emily Faithfull drew attention to Nightingale's work in the Crimea:

> I often hear of want of nerve, want of health, want of strength, want of judgement, till I wonder how such feeble creatures as women are supposed to be, ever contrive to live at all, and should begin to think of the story of Miss Nightingale and the Crimea as a fable, did I not know that some of the hardest work in the world is done by women, and under circumstances calling for an immense amount of fortitude and endurance.[22]

This was a curious and risky argument in that Nightingale's subsequent invalidism could easily counter it. However, there was also a certain security that any public criticism of Nightingale, no matter how appropriate, was quite beyond the pale.

The potential field of work for women doctors was often represented in the same way as the new hospital nursing, as an extension of women's philanthropic mission and their responsibilities within sanitary reform. Although the following passage published in the feminist *Victoria Magazine* in 1864 is from an article called 'Lady Doctors', it could quite easily be from any number of publications in the 1860s which promoted lady-nurses:

> [G]reat advantage ... may be obtained from the employment of really scientifically educated female medical practitioners among the poor ... Experienced and scientifically instructed ladies, acting as dispensary officers, would ... be instrumental in bringing about a

much better system of sanitary domestic management among our poor, and also frequently be the medium between the rich and the poor in charitable donations and assistance.[23]

Charles Drysedale, lecturer at the Female Medical College, wrote: 'Medical missionaries...are truly wanting in the poor quarters of all our large towns: but they must be ladies, and medical ladies too.'[24] The language of mission and philanthropy was one way for middle-class women to begin to resolve the contradictory demands of femininity with a yearning for independence and interesting professional work. Criticism of women's financial independence and career pursuits could be deflected by this philanthropic/mission contextualisation of their work. However, many of these women were caught in the contradictory position of both affirming their femininity, respectability and class status through this image of benevolence on the one hand, and on the other, actually wanting their need for, and right to, remunerative employment to be recognised. Some women found the emphasis on benevolent voluntary work constraining. Helena Pauline Dowling, student at the Female Medical College in the mid 1860s, wrote:

> I do not see why we are to be expected to attend the poor for small fees...there are a great number of unemployed women, and it is not always from motives of benevolence or charity that we are striving hard to enter the medical profession. We want to obtain an independent means of livelihood.[25]

The imperative to find employment for middle-class single women, an argument which the Female Medical Society in particular wrote into the discussion on women doctors in the same way it had been for nurses, cut right across the philanthropic construction of women doctors.

Despite the material and discursive connections between women's work as nurses and doctors, when it came to the treatment of men very clear distinctions were drawn. While it was accepted that women nurse men, it was unacceptable that women attend them as doctors. This was by no means construed as a limitation by women, as they had a very positive understanding of their role as physicians to other women and children. Even so, it is clear that these were the terms on which women were 'allowed' to be physicians. Women doctors were not culturally permitted to deal with male bodies. The issue of contact with men's bodies highlighted the very different meanings attributed to nurse–patient interaction on the one hand, and doctor–patient interaction on the other, even when the practitioner was female. So loaded was the doctor–patient relationship in subject–object and active–passive terms, that the very idea of women doctors implied a quite intolerable objectification of men. Because the subject position of the new female nurse came to be so understood through ideas of self-sacrifice and self-effacement, it posed no such threat. It was quite possible for women as nurses to act scientifically and therapeutically on passive male patients without objectifying them.

In English military hospitals in the 1850s and 1860s female nurses took over the care of men from male nurses. In these and other institutions, nurses were permitted very close bodily contact with (working-class) men, seemingly without upsetting either men's or women's sensibilities. For example, the Nightingale Training School records indicate that even the lady-probationers administered enemas to men.[26] Nurses sometimes attended and observed surgery, usually as attendants, but in certain cases more or less as students. This was especially the case around the 1860s and 1870s in England, before any real difference between the new 'lady-nurses' and the new women doctors had been established. Lucy Osburn spoke of surgeons at the King's College Hospital taking a special interest in her anatomical education in preparation for her position in Sydney:

> I knew some of the doctors personally, Mr Solly and Mr Geo. Clarke, and they always sent for me whenever any operations were to be performed, so that I had special opportunities of study in the operation-room... They would show me everything, point out the parts and organs of the body to me, and draw special attention to important matters, as they thought that such instruction would be useful to me.[27]

Much was made, however, of the consequences to women doctors if they had close bodily contact with male patients. The contradictions and inconsistencies between what was acceptable work for nurses and for women doctors were glaring, and were not lost on contemporaries. Elizabeth Blackwell wrote: 'At one time, women are ordered to keep their place, while at another they are assured that their place is at the bedside of the sick. Those who are most anxious to see women waiting upon male patients as nurses consider it an outrage upon propriety that they should attend their own sex as physicians.'[28]

As discussed in Chapter 4, the prospect of conventionally trained and regulated female practitioners gave a new focus to the sexed and sexualised nature of male midwifery and gynaecological practice. Much was made in these debates of women's reticence and modesty over vaginal examination by men, sometimes constructed in dominant discourse as a worthy modesty, sometimes as an irrational and unnecessary modesty, in which women failed to recognise the disinterested position of men of science. However, the prospect of women doctors highlighted an anxiety and a 'modesty' surely far more intense: the real taboo item was the penis. As nurses, women were permitted (therapeutic) contact with men. Yet acceptable bodily contact with male patients stopped short of contact with genitals. In the 1860s nurses at St Thomas's Hospital did not apply urethral catheters to men, although it seems that the catheterisation of women was fairly common.[29] The horror at the thought of women treating men's genital diseases was almost unspeakable, but produced the occasional comment in the medical press. Of two women studying medicine in Vienna, readers were informed that one was 'a most industrious attendant upon the syphilitic department of the hospital, the other, who is ambitious of becoming an operator, recently assisted in a case of castration'.[30] The somewhat frantic imperative to locate women doctors in the context of the care of women's bodies is certainly telling of women's inability and unwillingness to think of themselves treating men. But it can also be read as men's utter incapacity to contemplate medical genital examination by women and the objectification this would entail.

FEMINISM AND MEDICINE

Simply by being female and medical practitioners at the same time, women doctors disrupted the gendered order which structured 'professional' medicine. But as the cases of Mary Walker, James Barry, Elizabeth Garrett and Elizabeth Blackwell exemplify, there were various manifestations of this disruption. There were any number of ways in which each woman's femininity was paraded, hidden, rendered suspect and confronting or conventional and reassuring, by the women themselves, by their supporters and by their opponents. Women in the nineteenth century challenged the masculinity of conventional medicine in some cases by seeking to practice as men did, and in other cases by explicitly questioning the masculine terms through which medicine and health care was thought about. Some women aimed to make their sex irrelevant, while others saw their sex as precisely the issue, formulating an oppositional feminine medicine. The historian Regina Morantz-Sanchez has dealt at length with the different outlooks of nineteenth-century women doctors in the American context, comparing Mary Putnam Jacobi's support for reductionist and 'scientific' medicine with Elizabeth Blackwell's moral and feminine conception of medicine. Since the publication of her book, critically arranged around the gendered dualism 'sympathy and science', she has reread Elizabeth Blackwell's ideas in the light of recent critical theory and a late twentieth-century feminism of difference.[31] More sympathetic in this rewriting, she looks again at the feminist possibilities of Blackwell's insistence on keeping the moral and the spiritual within a feminine domain of medicine. What is hinted at in this later essay, but never quite addressed directly, is a body of scholarship on the history of feminism itself. In the British context, this history of feminism is entirely necessary to fully appreciate the conditions which created the possibility for an oppositional feminine medicine and to understand the feminist edge which the espousal of seemingly traditional femininity could acquire.

In Chapter 1, I discussed the considerable authority which middle- and upper-class women were able to claim from the cultural space of domesticity. Here, with reference to work on the history of feminism, I explore the implications of this more fully in terms of the development of a gendered politics of health, and the connections between medicine and Victorian feminism. Working through the meaning of Victorian ideas about 'separate spheres' has been the mainstay of the history of British feminism. Early scholarship identified the construction of the gendered spheres of the private and public, arguing that Victorian feminists aimed to 'escape' domesticity and enter the masculine public sphere as equals. Later work has explored the ways in which feminists sought not to eliminate a notion of separate spheres, but precisely to politicise the private and the domestic.[32] At a time when the meanings of femininity were in considerable flux, women cast about for possible redefinitions of the private sphere rather than ways to radically critique or reject it. At the risk of generalising about a remarkably diverse movement, ideas about an expanding woman's sphere were seen to be central to social progress. This facilitated the contesting of public space, a questioning of the nature of femininity and masculinity, and an easing of constraints on women's employment, education and location in the body politic. A strong belief in, and activism around, the idea of a 'woman's sphere' largely produced Victorian feminism. Not the only, but arguably the dominant Victorian feminist sensibility then was not one which saw sex as irrelevant, but one which saw sexual difference as precisely the political point. At every turn in the nineteenth century, discourses of femininity and domesticity are identifiable which not only oppressed women, but which were also shaped into a feminist challenge. As Barbara Caine has written: 'the very articulation of... feminism involved negotiating with and reworking Victorian domestic ideology... Had these mid-Victorian feminists not accepted and addressed the ideal of womanhood articulated in Victorian domestic ideology, they would not have been able to speak to their contemporaries at all. Once they addressed it, it was inevitable that the moral overtones of this ideal would become centrally involved in their feminist discourse'.[33] Paradoxically, the domestic and even philanthropic ideal of femininity was central to the formation of middle-class women's own political subjectivities, and to the development of an organised feminism. One site where this gendered politics manifested itself was in the domain of medicine and health.

Women like Florence Nightingale and Elizabeth Blackwell unquestionably formulated a feminine medicine in explicit and openly articulated opposition to mainstream medicine. In doing so, they firmly reinscribed 'woman' and particular women with traditionally feminine values and concepts: nature, healing, caring, nurturing. This is why Nightingale's ideas about women and gender have so often been assessed as conservative. But the corollary to this was the critical identification of mainstream medicine as masculine.[34] Such women's capacity to see the domain of health-care as gendered, opened up a crucial space for the questioning of dominant practices and attitudes. Some of the women who developed this critique of medicine were centrally involved in mid nineteenth-century feminism. Frances Power Cobbe and Josephine Butler are two of the most important examples. Others, like Nightingale herself, kept at a certain distance from the women's movement but nonetheless worked, wrote and thought in the same domain and with very similar notions about women's work and the questionable morality of the masculine medical profession.

Around the 1860s when debates about female medical practitioners and the place of women in medicine assumed some public prominence, Nightingale emerged as a firm opponent of women doctors. Elizabeth Blackwell wrote of Nightingale:

she is not prepared to endorse fully the medical idea, she is not unfriendly to it, but she has not realized the importance of opening medicine to women generally. This proceeds partly from her utter faithlessness in medicine – she believes that hygiene and nursing are the only valuable things for sickness, that the physician's action is only injurious, counteracting the useful efforts of nature.[35]

Nightingale was not opposed to women doctors on the grounds that women should not interfere in a male domain, but that she had little respect for that domain and the direction it was taking. 'I wish to see as few Doctors, either male or female, as possible', she wrote to John Stuart Mill. 'For, mark you, the women have made no improvement; they have only tried to be "Men", and they have only succeeded in being third rate men. They will not fail in getting their own livelihood, but they fail in doing good and improving therapeutics.'[36] She held a quite different view from other public women in the 1860s who were pursuing the issue of medical women vehemently, and there were frequent clashes. For example, Nightingale refused to sign the petition against the Contagious Diseases Acts because Josephine Butler had tied the issue to the need for women doctors: 'Mrs Butler could not ask me to sign her Petition for the repeal of the "C.D. Acts" without inserting a passage about the usefulness of Medical Women...However, I refused to sign till the "interlude" was taken out...*I* could not enter into the controversy without attacking the Medical Education of *men* – And this is impossible to me.'[37] While Nightingale eventually came to support the idea of women doctors, particularly in India, and also to support Elizabeth Garrett's New Hospital for Women, in the 1860s her opposition to Garrett's medical pursuits was expressed forcefully and regularly. But her most fundamental criticism was always saved for the male medical profession:

> She [Garrett] starts on the ground that the summum bonum for women is to be able to obtain the same Licence or diploma as men for medical practice.
> Now I start from exactly the opposite ground.
> Medical education is about as bad as it possibly can be.
> It makes men prigs.
> It prevents any wise, any philosophical, any practical view of health & disease.... Miss Garrett does not say... how can we give women the best general Medical Education? She says:-how can we satisfy

the 'Examining Board'?...'Examining Boards' are just so many charlatans.[38]

Nightingale wished to distinguish the training of nurses and midwives from any sort of medical training, and discouraged 'ambitious Lady Doctors in embryo' from entering the training scheme at St Thomas's.[39] For Nightingale, a female medical practitioner would ideally be a midwife/nurse, yet in Nightingale's usage, midwifery included the treatment of the diseases of women and children in addition to attendance at childbirth.[40] Like Blackwell, the model of La Maternité in Paris, in which women were trained at the bedside and in lectures largely by 'Lady Professors', impressed her greatly.[41]

Nightingale's image of a female medical practitioner was of one who implemented a health-care practice which was different in kind from scientific, masculinist, interventionist medicine. As Sandra Holton has argued, Nightingale was fundamentally critical of medical efforts to control disease rather than to promote health, and saw the promotion of health clearly as a feminine enterprise.[42] She thought of health explicitly in terms of 'Nature the healer' not 'Man the curer': 'Nursing is putting us in the best possible conditions for Nature to restore or to preserve health – to prevent or to cure disease... Sickness or disease is Nature's way of getting rid of the effects of conditions which have interfered with health. It is Nature's attempt to cure – we have to help her. Partly, perhaps mainly, upon nursing must depend whether Nature succeeds or fails in her attempt to cure by sickness... For it is Nature that cures: not the physician or nurse.'[43] Of course, this usage of 'nature' was a thoroughly gendered one. 'Nature' was almost exclusively personified as feminine, and women were discursively linked to the concept of nature in a multitude of ways.[44] Women's use of 'Nature' often functioned as a potent symbol of innocence, especially when posited against science and technology.[45] Nature and science were binarised and implicitly gendered such that a close affinity with masculine Science on the part of women necessarily undermined their feminine Natural sympathies. Women were seen to work with nature, men against it (or 'her'); a binarised notion not easily overturned, indeed quite central to one strand of late twentieth-century feminist thought.[46] Female physicians, wrote one commentator, are 'the handmaids of nature, possessing all the qualities for good nursing'. They would keep a check on invasive medicine, and would press for 'milder methods of treatment...[and] co-operation with nature.'[47] Female medical practitioners were often thought of as benign healers, a position increasingly represented by the newly feminised, moralised figure of the nurse.

Men and women, feminists and anti-feminists, all thought about 'the woman question' in some way within the discourse of domesticity. However, the distinction between a feminist use of the discourse of domesticity and a reactionary use was often fragile and precarious. The distinction between Nightingale and Benjamin Ward Richardson's notions of women's place in the world of medicine and health care is a case in point. Like Nightingale, Richardson held unfavourable opinions about interventionist and reductionist medicine, and the two influential figures were in broad agreement about women's place as sanitary reformers. They both presented the pursuit of hygiene or health as a feminine practice, and the practice of actively intervening in sickness as masculine: Hygeia was the daughter of Aesculapius. Like the contributor to the feminist *Alexandra Magazine*, Richardson was at pains to connect female practitioners with a domestic nursing/medical practice:

> [D]oes she [woman] not possess a true instinct for medicine? Who should doubt the existence of such an instinct in the mother caring for her children? The 'Epsom Salts' and 'Senna-Tea' dear to the infantile memories of every true Englishman, and sacred to the domestic hearth of every true English home, are they not, indeed, true symbols of this instinct?

Like Nightingale Richardson wanted 'female physician' to mean midwife:

> It is highly probable that there is a noble field open to female enterprise, and one singularly well adapted for the highest display of feminine talent, in midwifery...If we must have female physicians...for the credit of womankind, they should first devote themselves to the acquisition and practice of that branch of medicine in which the aid of man is most obnoxious, and is tolerated, perhaps, more from necessity than choice.[48]

While Nightingale's ultimate vision of health involved an independent female sphere, Richardson's vision was about containing women's intervention within an appropriate sphere defined by its gendered opposite: women were not to become doctors. Richardson's enthusiasm for 'women as sanitary reformers' certainly contributed to the opening of some opportunities for women, yet like the Female Medical Society, it also needs to be assessed as a response to the challenge then being posed by Elizabeth Garrett, Sophia Jex-Blake and others. His later writing makes this clear: 'There are women who think it the height of human ambition to be considered curers of human maladies... I would with all my strength suggest to women that, to be the practitioners of the *preventive art of medicine*; to hold in their hands the key to health... would... be... much above the exercise of curative art... I press this office for the prevention of disease on womankind, not simply because they can carry it out... but because it is an office which man never can carry out.'[49] Even more defensively, Richardson argued that the 'female sanitary scholar' should not be educated in diagnosis as 'the art is not necessary for women except in a limited degree'.[50]

Richardson's was a very subtle argument against women doctors and against women's autonomy as practitioners. It gently blunted the feminist edge of the discourse of domesticity through a re-appropriation of that discourse. Other arguments were not so subtle. The following is a very shrill Australian medical response to the English debate on women doctors in the mid-1860s, and shows not only how unthinkable women's curative contact with men was but also how the whole domestic and feminine construction of women doctors could easily be inverted into weapons of opposition. After praising the work of Nightingale, the writer went on to berate the idea of women as doctors:

A great deal of flatulent eloquence has been expended by the champions of so called women's rights, in an endeavour to prove that the healing art is peculiarly fitted to display the qualities of benevolence and kindliness, essentially belonging to the female character... The sick chamber is desolate if there be no women in it... But a woman who dissects, who makes post mortem examinations, who tests urine... who can pass the male catheter... who punctures buboes, probes sinuses, examines dejecta, sputa and purulent discharges, applies ligatures to haemorrhoids, and may have just come from operating for fistula in ano, is not a person in whom you would look for the tenderer domestic qualities.[51]

It is interesting to note the way in which doctors are depicted in this passage as dealing with bits of bodies. By reducing medicine into a practice concerned with the fragmented component parts of sick people, women were effectively excluded. Such reductionist medicine offered no space for the play of morality or spirituality; there is no soul to work with if one's practice is about urine, fistulas and sputa. It suggests that this type of practice was precisely what medicine increasingly saw about itself as most scientific, most masculine and most prestigious.

Knowledge grounded in the scientific method and in rational understanding of the material in ever-reduced units enjoyed a new status in the late nineteenth and early twentieth century. Modernity was marked by a discursive prising apart of the physiological body from its moral or spiritual meanings. According to Jean Comaroff, 'biomedicine has become progressively disengaged from the language of cosmology and morality'. This involved a process of 'reification, stripping away the social and environmental underpinnings of disease'.[52] By the early twentieth century it was possible to think of the modern body as a concrete, physiological entity, reduced and separate from a moral, mental or spiritual life. Of course, this way of thinking about the body in health and illness was as constructed and culturally produced as any other. 'Godliness', or metaphysical meanings of the body, did not drop out of the equation but were masked by this modern rationalist discourse. Rather than disappearing, cultural meanings of the body shifted to a range of peculiarly early twentieth-century concerns: national health, sexual hygiene, racial hygiene and so forth. It is no coincidence that women doctors were to be found everywhere in the production and practice of this slightly different field of 'hygiene' in the early twentieth century.

It was this reductionist vision of illness and healthiness, or more precisely its contagionist antecedent, which Rosenberg has argued was so unthinkable to Nightingale.[53] Once again, the gendered dimensions to this issue need emphasising, for the rejection of the reification of body, the insistence that the physical body hold some sort of metaphysical meaning, was an issue central to the thinking of many Victorian feminists. Current feminist interest in ideas about a mind/body dualism turns out to have a long history indeed. For example, in her numerous essays on medicine, Frances Power Cobbe was insistent that the capacity to understand the body as separate from the soul or mind, then being formulated by modern medicine, was to be challenged and overturned: 'There is, of course, a great and ever-present temptation to a physician to view things from the material, or (as our fathers would have called it) the carnal side; to think always of the influence of the body on the mind, rather than of the mind on the body... to study physiological rather than psychological phenomena.'[54] For her, the mind and the soul must continue to be seen as an inextricable part of the physical body. Cobbe dealt directly and critically with questions of the secularisation of medicine which accompanied and produced this reification of the body: 'Health of body has been accorded the importance which the – real or supposed – interests of the soul alone commanded two centuries ago'.[55] Certainly these types of issues exercised sanitarians as discussed in Chapter 1 but Cobbe's analysis, amongst others, highlighted and pursued the specifically gendered dimensions to such questions.

The question of morality being written out of medical science was constituted as a feminist one because middle-class women had all kinds of investments in their own individual and collective moral status. Quite simply, feminists (and others) understood women to be morally and spiritually superior to men. This firm belief in the social responsibility and capacity of women to improve the moral lives of men and the nation – in the material interests of women, amongst other things – unquestionably shaped much of Victorian and Edwardian feminism. That 'purity' which had long been inscribed on woman, was reworked and capitalised on by feminists. Social purity and indeed sexual purity came to be a political and politicising issue for many women, as they struggled to 'raise' men to their moral standard, in the specific and demonstrable interests of women. Purity was conceptualised as something of a base for women's power in social and political battles, from the Contagious Diseases Acts to temperance to the role of women in the citizenry. Many feminists saw their place as purifiers of a contaminated and corrupt public world.[56] Importantly, arguments about women's purity and power were often located within a discourse of health, disease and bodies, specifically setting up a spiritual/sexual purity of women against a non-spiritual, 'brutal' corporeality of men. For example, one major site of such feminist argument was debate on venereal disease in which women, seen as pure and innocent were contaminated, in both moral and physical senses, by an abusive sexual double standard which worked in the interests of their husbands. The mission of feminism was to purify such men, to raise them to women's standard and to abolish prostitution. Many women doctors worked, wrote and campaigned around such issues. Feminism and the domain of women in medicine shared a language of purity and hygiene, a language which always sat between and spoke to both moral and physical concerns.

Elizabeth Blackwell was without doubt the foremost figure in promoting and practising a type of 'feminine medicine' which stemmed from such convictions about the different natures of men and women. She wrote and argued fervently that morality and spirituality needed to be central to medicine as a philosophy and a practice. Women's maternal powers, instincts and natures, as they were (problematically) constructed by Victorian feminism, were brought to bear on this issue. Blackwell wrote of women in medicine:

> It is not blind imitation of men, nor thoughtless acceptance of whatever may be taught by them that is required, for this would be to endorse the widespread error that the race is men... Now, the great essential fact of women's nature is the spiritual power of maternity... The legitimate study and practice of medicine... requires the preservation in full force of those beneficent moral qualities – tenderness, sympathy, guardianship – which form an indispensable spiritual element of maternity... The true physician must possess the essential qualities of maternity.[57]

Along with Cobbe she argued that '[o]ur ministrations to body and soul cannot be separated by a sharply-defined line'.[58] By the last decade of the century, even as such sensibilities were gaining a feminist meaning, they were increasingly marginalised within mainstream medicine and science. There is a certain desperation in many of Blackwell's tracts and speeches written around this time. She realised her position was an embattled one, to say the least. In *Christianity in Medicine* (1890) she admitted that '[t]he expression "Christian Physiology" which I deliberately use as designating really vital truth in relation to human nature and to progress in the Healing Art, calls forth the sneering or hostile criticisms of many'.[59] Neither in her medical philosophy nor her religion (she became a Christian Socialist and later a Theosophist) was Blackwell orthodox.

While Blackwell should be seen, as she came to see herself, on the fringes of late nineteenth-century medical thought, there were identifiable ways in which the careers of many women doctors did bear out her notions of a feminine medicine, even if these women themselves would understand their work in different terms. Blackwell wrote in 1889: 'There are two great branches of medicine whose importance will, I hope, more and more engage the attention of women physicians. These are midwifery, which introduces us to the precious position of the family physician; and sanitary or preventive medicine, which enables us to educate a healthy generation.'[60] While there were certainly some female surgeons and bacteriologists over this period, a disproportionate number of women did find themselves in fields like public health and preventive medicine, missionary medicine, and (especially in the early and mid twentieth century) the enormous growth industry of maternal and infant welfare. The idea of 'women as sanitary reformers' held considerable currency well into the twentieth century. Blackwell herself formed the National Health Society which promoted preventive health and education in hygiene, a project which she called her 'serious life work'.[61] Mary Scharlieb, to take another example, was especially interested in what she called the 'socio-medical work' of women doctors, again organised around the concepts of public health and preventive medicine with an early twentieth century nationalist inflection: 'one of a woman doctor's privileges and duties is to assist in preserving the health of the nation.'[62] Such ideas dovetailed quite neatly with early twentieth-century eugenics, that other social movement which revolved around a notion of 'purity'. For example, Louisa Martindale, another woman doctor centrally involved in a series of feminist campaigns wrote:

'[Woman] is the creature to whom the Race is more than the Individual, the being to whom the Future is greater that the Present', and perhaps it is because of this inherent, deep-rooted, ineradicable quality of visionary foresight that no work attracts some women more than that of a life service in the cause of eugenics or Research.[63]

In the mid and late nineteenth century there were strong material and discursive links between women doctors and nurses, and between these and the culture and practices of women's sanitary reform examined in Chapter 1. The more that familial and domestic contexts were foregrounded, the more women could locate themselves with authority in the discourse as nurses-cum-doctors. There was an attempt to make discursive space for women, by shifting the understanding of medicine itself, by maximising the depiction of domestic and philanthropic work, and by minimising the depiction of scientific, interventionist or experimental medicine. In the modernisation of both nursing and doctoring for women, feminine meanings were mobilised and the possibility of feminine subjectivities was sustained, largely through the domestic and moral imperatives of sanitary reform and the notion of philanthropic work. Sometimes such a discourse managed to assume a confronting and effective feminist meaning, but often enough it served to reinscribe women within a set of traditional and constraining values. Nineteenth-century feminist renegotiations of domesticity and femininity undoubtedly reinforced a set of gendered binaries which postmodern feminists seek to radically problematise and collapse. In this sense, recent feminist theorising has embarked on a long-overdue and most welcome paradigm shift in thinking about gender. But these binaries still need to be understood, analysed and interpreted historically as fundamentally structuring nineteenth-century culture, including the culture of nineteenth-century feminism itself. While James Barry is a fascinating case (who deserves further historical study) in the radical scrambling of gendered identity for her own purposes, the feminine medicine articulated by Elizabeth Blackwell and others was a strategy far more familiar, possible and achievable for most women in the field of health. In all of these ways these women insisted on, and negotiated for themselves a gendered politics of health.

6 Dissecting the Feminine: Women Doctors and Dead Bodies in the Late Nineteenth Century

For all the scholarly interest which 'the body' has elicited in recent years, the bodies under question have usually been those of living human subjects. The dead body, which (to state the obvious) has no subjectivity, consciousness, agency, or volition, has not been quite so interesting for those involved in the postmodern project of problematising the mind/body distinction. But the human corpse is a cultural and historical product like any other version of the human body, even if it is not a 'subject' as such. Given the proliferation of discourses which produce 'death' in any given culture, the dead body turns out to be a very inscribed body. Part of my objective in this chapter is to examine the discursive construction of the dead body at a particular historical moment and cultural site: the late nineteenth-century dissection room. What I am especially interested in is the idea that dead bodies were still sexed bodies, and that particular discourses of gender and sexuality shaped the encounters between students of medicine – male and female – and what they called their 'subjects'. Specifically, I focus on the sudden foregrounding of such discourses of gender and sexuality prompted by the entry of women into conventional medical education; women who, if they wanted to register as a doctor, had to dissect the human corpse several times over. The question of women cutting, opening and investigating the insides of the human body compounded what was already a deep cultural anxiety about the practice. Some women doctors as well as their proponents argued against women's place in the dissection room and by implication in the operating theatre: indeed in any practice which involved the use of knives. Other female medical students vehemently opposed the idea of dissecting with men, although they were comfortable enough doing it alone. And it was common indeed for male students and their anatomy teachers to protest in the strongest possible terms the presence of women in the dissection room. The point is not that women did not practice dissection (or surgery), for they did. Rather, it is the constant discussion about the issue, the terms of this discussion and the anxiety-producing cultural implications of bringing the dead body and the female body together, which I analyse here. What was so unsettling about the contemplation of women cutting up bodies? After all it was women who dealt with the dead all the time in the domestic context, preparing the corpse for burial. And in what precise ways was the practice of dissection gendered? Why could not women wield the knives?

Ludmilla Jordanova has identified the practice of dissection as a strongly sexualised one.[1] Her analyses of visual and literary representations of the female corpse, and the construction of the process of dissection as penetrative, as a way of 'knowing' women, demonstrate the range of ways in which viewing the dead body and cutting it up were constituted in terms of an active male subject and a quintessentially passive female–object–body. While her work provides a rich starting point for thinking about the dissection room as a gendered and sexualised cultural space, I want to turn to the specific problematisation of the woman-as-subject in this encounter. In the figure of the woman-dissector, Victorian anxieties about both transgressive women and the opening-up of the dead body became particularly acute. Moreover, these women disrupted the gendered and sexualised ways in which dissection was imagined. For while the corpse of art, literature and the popular imagination was young, female and sexualised, in the actual dissection rooms and anatomy textbooks of the late nineteenth century, the corpse was nearly always male.

Women's concern about, and sometimes implacable opposition to, dissecting with men was not simply a manifestation of nineteenth-century feminine modesty. It was about the difficulty with which the male-subject/female–object dynamic was reversed. It was also about a nineteenth-century feminist discourse which politicised bodily boundaries and which rendered the idea of 'going inside' a body – be that through the practice of medical examination, vivisection of animals, surgery or dissection of humans – somehow an invasive, violent and masculine practice. And as far as men's difficulty with female colleagues in the dissection room is concerned, this was not simply an aspect of masculine protection of their own education, knowledge and space, although it was clearly each of these. Women-dissectors – female bodies marked as disordered – destabilised what was already a profoundly unstable and culturally precarious practice. Notwithstanding the objectivist scientific posturing of nineteenth-century anatomists and medical students, men cutting up human bodies worked in decidedly dangerous margins between life and death, inside and outside the body.

In thinking through the peculiar difficulty and debate elicited by women dissecting, I have utilised not only the work of cultural and medical historians, but also theorists who have been intrigued by the battles which take place at boundaries – especially at bodily boundaries – and by the metaphoricity of the female body in such battles. In particular, Julia Kristeva's concept of the abject offers one productive way to read the dense cluster of boundaries and bodies at issue in the late nineteenth-century dissection room.[2] The abject is that which needs to be expelled in order for the 'clean and proper' body, 'clean and proper' subjectivity, to be asserted. Further, Kristeva suggests that the abject can never be fully expelled, but that it lingers on thesholds of all kinds, neither subject nor object. Despite massive cultural efforts to render the corpse object, the lingering anxieties about dissection suggest the impossibility of doing so; the corpse, especially when dismembered and disintegrating, was always abject.

CONTEXTUALISING DISSECTION

In the mid and late nineteenth century, Western culture generally was very unsure about dissection. On the one hand, there was a marked popular unease about the scientific trade in corpses and about ethical and religious issues of the mutilation of bodies. On the other hand, dissection was a practice becoming inseparable from a new type of scientific medical education, itself gaining increasing status and recognition as the century progressed. In her wonderful book *Death, Dissection and the Destitute* Ruth Richardson has analysed the profoundly negative cultural understandings of dissection in late eighteenth- and early nineteenth-century Britain.[3] Dissection was closely connected with punishment, a legacy of the longstanding Western practice of public disembowelling and flaying. Hanging and dissection was the most severe sentence in the eighteenth century; the ultimate corporal punishment. For a considerable time medical men, scientists and anatomists received their bodies from the gallows. Richardson examines the subsequent development of a trade in corpses – body snatching – generated by an increased demand on the part of doctors, new anatomy schools and other institutions of medical education.

This illicit trade implicated doctors and anatomists in a criminal and immoral practice, one which prompted extraordinary popular objection as the many riots and effigy-burnings of dissectors testify. As a way of stemming the illicit trade, yet ensuring a supply of bodies, the Anatomy Act of 1832 allowed for doctors to gain their bodies from those who died in government workhouses and hospitals, and this remained the main source of bodies over the century. Richardson argues that this move engendered a deep fear of hospitals and of 'dying-on-the-parish,' identifiable in working-class and popular culture well into the twentieth century.[4] 'Dissection,' she writes 'represented a gross assault upon the integrity and identity of the body *and* upon the repose of the soul.'[5] In these understandings, the dead body was never simply generic human flesh, but was always a particular dead person.

At the same time, dissection was becoming an entrenched part of a new institutional medical education. Dissecting the human body was the accepted means by which medical students familiarised themselves with human anatomy, and it was seen specifically to be a training for future surgical practice. From the middle of the nineteenth century, dissecting dead bodies and surgically operating on live ones were coming to be thought about similarly, in the sense that anaesthetics now rendered the live subject unconscious – they were temporarily 'dead'. The cultural spaces in which dissection and surgery occurred also drew meaning from each other, both as spectacles, as suggested by the term operating 'theatre', and as privileged masculine sites where the gathering of special knowledge from inside the body took place. Consider these instructions to medical students:

> Each student with his own hands dismembers or dissects the body, so as to see and feel each constituent part of it, recognise its characters and learn its relations to the surrounding structures. Moreover, he is here trained to handle his instruments, and so is fitted for the practical work of his after life.[6]

Such descriptions of dissection suggest the sense in which the practice was a manual one. In this period, medical practice generally was corporeal and physical work, as much as, and perhaps more than it was intellectual work abstracted in books and lecture theatres. Far from being an intellectual procedure, dissection involved the body of the dissector as much as the corpse. Dissection was about two bodies, not one – and two sexed bodies at that.

Over the nineteenth century, such 'gross' anatomy was refined and augmented by knowledges produced by more sophisticated technologies such as the microscope and by laboratory sciences.[7] Dissection/anatomy came to be less a high-status specialty than a grounding knowledge, a practice associated particularly with students and with medical education. One anatomist described his work in very mundane terms:

> While the anatomist limits himself to describing the form and position of organs as they appear exposed, layer after layer, by his dissecting instruments, he does not pretend to soar any higher in the region of science than the humble level of other mechanical arts, which merely appreciate the fitting arrangement of things relative to one another and combinative for the particular design of the form, of whatever species this may be, whether organic or inorganic – a man or a machine.[8]

Such anatomical dissection was distinct from an autopsy or post-mortem, which was undertaken by qualified surgeons and physicians usually within a hospital setting, rather than at a medical school or university. Dissection assumed its meaning within a pedagogical context. Whereas dissection aimed to educate in normal human anatomy, the post mortem aimed to reveal knowledge about a body in some way abnormal, sick, diseased. It is telling of nineteenth-century conceptualisations of sexed bodies that dissection almost always involved the male body (that is, the normal or standard human body) while autopsies often involved the female body (the pathologised, abnormal body).

In large part, tracing changing cultural attitudes toward dissection is about tracing changing imaginings of the body. Cultural acceptance and approval of scientific dissection required an understanding of the dead body as soul-less, as inert, as flesh only, as object without any subjectivity. A reductionist medical vision increasingly permitted such an understanding, a vision in which body parts were removed from one another, and where the body in death was reified, was divested of moral or spiritual significance. But it took some time and a major secular shift for this to be the case. By the mid twentieth century, having one's body dissected had shifted from a sign of criminality, poverty and fear, to being a matter of 'giving one's body to science' as a goodwill gesture towards human progress. The period in question here, however, was one in which there was no sure conceptualisation of the dead body. There was no certainty of the scientific perception of the corpse as an objectified piece of flesh, a conceptualisation clearly necessary for dissecting a dead human with any peace of mind.

The dissection room and the unsettling practices which went on within it gave rise to stories which suggest this incapacity to thoroughly objectify the corpse. Some stories, such as an 1889 medical student's poem about being locked in the dissection room overnight and being himself dissected by the spirits of the corpses, articulated considerable anxiety about dismembering corpses.[9] More common stories – cavalier accounts of macabre pranks, student revelry and practical jokes in which students threw around dismembered limbs and suspended corpses from the ceiling – belied this anxiety. Surgeon John Bland-Sutton remembered the dissection room in terms that were explicit about the masculinity of the space: 'In my first year (1878) the dissecting-room was regarded more or less as a Sports Ground. On two occasions I disturbed a boxing-match, and a rat-hunt was another form of diversion.' He named the dissection room as a sort of second common room for students.[10] Yet these were as much stories as actualities, perhaps told and retold by way of trivialising what was the most disturbing aspect of these young people's entré into the world of medicine and the human body. When Elizabeth Garrett entered a dissection room for the first time in 1861 she wrote to Emily Davies: 'The reports have been gross exaggerations. There were no bodies hanging over chairs or by their feet from the ceiling.'[11]

Such stories and anxieties about dissection on the one hand and the scientific normality which it was gaining on the other, the general cultural and popular wariness about the practice, and its construction as a physical, sensual practice, rendered women's participation difficult: not impossible, or unthinkable, but difficult. The woman-dissector disrupted a set of expectations about the relationship between dissector and corpse, rendering it a major topic of discussion in the context of changing medical practice. In Edinburgh, the immediate issue which gave rise to the famous riot of male medical students in protest to female students, was the latters' participation in dissection in the company of men: 'the systematic infringement of the laws of decency by the dissection of female or male subjects by women, in the presence of men...we protest against the excessive indecency of young women and men studying anatomy and physiology in the same anatomical school'.[12] In other cases, the situation was reversed, and it was women medical students themselves who refused to undertake the practice with men. At the University of Melbourne in the late

1880s, for example, the question of a separate dissection room for women was raised over and over again, as the women insisted upon it, and the Faculty protested the expense and inconvenience. In 1887 the women wrote as a group: 'we are reluctantly compelled ... to express our intention of discontinuing our work here if no such concessions can be granted us'.[13] And in 1895: 'We are decidedly opposed to mixed dissecting, and we are sure our work would be greatly hampered thereby.'[14] Though played out differently, the question of women dissecting, and especially dissecting in the same space as men, was problematic and controversial. As always, nineteenth-century exclamations about the 'laws of decency' and the 'protection of modesty' unmistakably signalled some sort of threat to, or transgression of, dominant systems of gender and sexuality.

SEXUALISING DISSECTION: THE GAZE AND THE TOUCH

The domain of dissection was invested with sexual meanings partly through the longstanding Western cultural connection between death and desire, the morbid and the erotic. The historian Philippe Ariès has written extensively on the significance of eighteenth-century romanticisation/sexualisation of cemeteries and tombs; the necrophilia stories which run through de Sade's work; the imagery of male Death not simply taking a young girl's life, but raping her: 'Death no longer merely points out a woman, his victim, by approaching her and drawing her away by an act of the will; he violates her, he plunges his hand into her vagina.'[15] There was a perceptible undercurrent in which dissection was rendered a sexualised and violent practice; the indecency of the dissection room was in part a sexual indecency. Consider this similar story recounted in an 1877 text:

> the dead body of a well favoured girl of about 15 years was brought to St Bartholemews Hospital for dissection. Although no marks of violence were apparent, the students were of opinion that she had not met her end by fair means. One of these introduced his finger into the vagina, and finding the hymen to be intact, declared that she was a maid. Upon this the porter who was employed to carry the dead bodies in and out, also put his finger up, and exclaiming 'that he had never had a maidenhead, but that he would take one now, by G-,' proceeded to violate the corpse then and there, in the presence of the students assembled.[16]

Ludmilla Jordanova begins her book *Sexual Visions* with a quote from Turgenev's *Father and Sons*, in which the hero-scientist sees a beautiful woman, falls in love and exclaims: 'What a magnificent body! Shouldn't I like to see it on the dissecting table.'[17] This impossibly transparent piece of writing collapses the desire of the anatomist for the corpse into the desire of man for woman. *Sexual Visions* offers important ways of thinking through such a statement, examining the processes by which gender and sexuality shaped the culture and the discursive practices of medical science. Through the eighteenth and nineteenth centuries in the West, she argues, the cultural linking of woman with nature, also made woman the quintessential passive object of investigation by biomedical science; she explores images of a personified feminine Nature unveiled before a masculine Science, her secrets revealed. One site where this dynamic was reproduced with some regularity was in visual and literary images of the corpse – specifically female – dissected and penetrated by groups of scientific men. Often these representations rendered the female corpse unmistakably sexualised; adorned with jewellery, long flowing hair, breasts partially and seductively revealed and so forth. Such representations were a classic formulation of female powerlessness and male invasion; men's desire for the female body conflated with desire for knowledge.[18] This genre of representations of the young, beautiful, sexualised female corpse, surrendered to the male gaze and the male capacity to penetrate might also be seen as part of a peculiarly Victorian 'cult of the beautiful dead'. In her feminist psychoanalytic study *Over Her Dead Body* Elisabeth Bronfen writes of a whole literary and imaginative culture of male lovers gazing at feminine corpses – their lost lovers. She analyses a new sort of spectacle over the female corpse, over the dying woman in particular, and the cultural construction of a sentimentalised moment in which the dying/dead woman was an object of (sleeping) beauty.[19] There was, then, considerable cultural investment in a gendered and sexualised understanding of dissection in which the masculine scientist/dissector penetrated, came to 'know' the feminised corpse in a dynamic shot through with all types of desire.

The female dissector clearly disrupted this desire when she entered the dissection room. Not only did she disrupt it, she reversed the gendered subjectivities and the sexualised dynamics which operated there. If the dominant representation was of the male dissector and female corpse, the female student, more often than not dissected the male corpse. While there was a proliferation of literary and visual images of a sexualised female corpse, manuals of anatomy for medical students represented the male body almost exclusively.[20] For many men, encountering women investigating and dissecting the genitals of corpses was intolerable. It was certainly so for the participants in the Edinburgh riot against female medical students who, having tolerated the women for several years, saw the practical anatomy lessons which began in their third year as quite beyond the pale. Their riot was planned for the day when the perineum of the corpse was to be dissected. In letters to the more tabloid of the medical press, the students articulated the precise nature of their problem: 'the "lady students" have been dissecting the genital organs in the same room with the gentlemen students!'[21] Another wrote: 'by dissecting the human genital organs... the female students have unsexed themselves'.[22]

Equally, part of the problem for the woman-dissector herself was the issue of the male corpse. At one level women's concern about dissection, and their common insistence that they dissect separately from men, may have been about the nakedness of the corpses. The difficulty for women was not simply about visualising a dead male body, it was also about the gendered nature of the gaze through which this encounter took place. Surveying the female body – dead or alive – serves to construct the authority and control of the male subject. The corollary is, of course, that 'Woman doesn't look, she gives herself to be looked at.'[23] While numerous theorists of the gaze have argued that in fact the female subject also looks, indeed looks back, and is never simply looked *at*, in the dissection room this is not the case. In the dissection room, the dead body never looks back; the gaze is only ever one way. In this sense, the gaze of the dissector *is* the male gaze *par excellence*. The scientist, the dissector, can ponder the body as long as he wants, without ever having to negotiate a returned look.

The images of dissection reproduced in Jordanova's book illustrate this perfectly. *Der Anatom* (1869) by Gabriel Max reveals an intimate moment in which the anatomist's gaze lingers over the female body, a look which is suspended permanently. She is all his and he has complete power over her, power to contemplate her body as long as he (and the viewer of the painting) wants. In part then, it was the potency of the gaze in the dissection room, the thorough gendering of this gaze, which rendered it so difficult for women to position themselves as subjects. An encounter between female subject and male body threatened to reverse the symbolic economy of the gaze to an intolerable degree. It created a space where women tried to assume the position of looker, penetrator, active subject in which man was rendered the body, object, penetrated, looked at. But what discourse was available to enable women to conceptualise or actualise themselves as the penetrators, the operators, the dissectors, the knowers, the subjects with the desiring gaze? Put simply, women were to be the objects, not the subjects of science.

In this context, the constant discussion over mixed or separate classes can be understood. Left alone, women could possibly begin to conceptualise themselves as subjects, but with men in the same space any such possibility was undermined. Again and again, women complained of being watched by male students, of being ridiculed, of becoming themselves object. One wrote, for example: 'though we all attend the same lectures, we object to some of the work being done in the same room with the men. I refer to dissecting... What business have they to come in when we are dissecting? Criticising, of course, and even laughing at us.'[24]

If the gaze involved in dissection was a desiring one and a (hetero)-sexual one, and the bodies were often male, it is certainly worth asking how the male gaze on the male corpse was negotiated. Clearly men were not permitted to take too much pleasure in the objectification and penetration of other men, albeit dead ones. There were a number of strategies by which the sexed embodiment of both male dissector and corpse were disavowed, strategies which were by no means necessarily consistent with one another. Firstly, there was a sense in which the female corpse was marked as sexed, yet the male corpse could be seen as the sexless, generic, universal human body. Secondly, there was the effort to render the corpse not man but machine in which bits and pieces fitted together and were taken apart in particular ways. And thirdly, any illegitimate sexualisation of the penetrative encounter between male dissector and male body, could be partly negotiated by the posturing of these men-of science not as embodied subjects, but as 'mind'. One way of thinking about the encounter between male dissector and male corpse was as a scientific encounter between mind and body in its most extreme form; in this sense not to render the corpse sexless, but precisely to feminise it. And here it is worth noting how very feminised the male corpse did become at certain moments. A most striking example is the placing of the male body in the lithotomy position for dissection of the perineum: the position which doubled as the classic for heterosex and for gynaecological examination. While legs raised and separated in stirrups for the convenience of the dissector/surgeon/obstetrician was a disturbingly common image of the female body, it is rare indeed to find the male body represented in such an open, available and vulnerable way (see Figure 6.1).[25] A sexual discourse haunted men's penetration and dismemberment of other men's bodies. Even before entertaining the 'problems' brought by the woman-dissector, men in the dissection room were required to negotiate a very peculiar and precarious set of cultural–sexual meanings, and it seems to me that this compounded the general anxieties which dissection produced. There was a real uncertainty about how to think about the corpse – an imaging of the corpse as feminine, yet also as generic man/human; as sexualised, yet also as sexless machine. Both the representation of the male corpse as machine and the feminisation of the male corpse can be seen as strategies to objectify it.

It is clear enough that women also employed such strategies of disavowal. They articulated corpses as 'mummies' or as being 'not like human things at all'.[26] They also tried to distance themselves; to become a 'mind' at work. One woman wrote, in response to insinuations of her sensual/sexual enjoyment of dissection: 'The body, after the necessary preparations, resembles that of a mummy rather than one where life is recently extinct ... the mind is interested in the work, and not filled with sensual thought, as hinted.'[27] However, just as women had difficulty assuming the position of active subject, so with the position of 'scientific mind'. There was always a sense in which she was herself 'body,' inscribed by the same discourse of woman-as-body-as-nature which marked the corpse as feminine. If male anatomists and medical students tried desperately to figure the production of knowledge from dissection through the coming together of minds and bodies, women dissectors disrupted this by creating a symbolic encounter between two bodies. The difficulty for women was also because of the *particular* meanings which cohered around the female body in the nineteenth century. The metaphoricity of the female body specifically is at issue here, as it was for the significance of the old nurse and the new nurse. Elizabeth Grosz suggests that in the West, women's bodies have been constructed as 'leaking, uncontrollable, seeping liquid ... a disorder that threatens all order':

> The metaphorics of uncontrollability, the ambivalence between desperate fatal attraction and strong revulsion, the deep-seated fear of absorption, the association of femininity with contagion and disorder, the undecidability of the limits of the female body ... are all common themes in literary and cultural representation of women.[28]

Figure 6.1 Dissection of the Male Perineum

Source: D. J. Cunningham, *Manual of Practical Anatomy*, 1893.

Women in the dissection room, by their very femaleness, heightened considerations of the fleshliness and corporeality of the dissector, which scientists tried to disavow.

But I would like to refine and historicise Grosz's formulation of the cultural meanings of the female body in terms of its particular nineteenth-century manifestations. As we have seen, inasmuch as the female body meant 'pollution', it also meant 'purity'. The two did not contradict each other so much as demonstrate the intensity of anxiety about the female body, always overinvested with meaning. Dichotomous constructions of 'woman' as pure and contaminated certainly informed understanding of women doctors, as it did most profoundly of the new chaste pure female nurses. The language within which dissection (and surgery) were discussed was sexual and bodily, a language structured through powerful dichotomous concepts of dirt and cleanliness, filthiness and sterility, purity and impurity. As I have suggested in previous chapters, this discourse was entirely commensurable with ideas about the female body specifically, rather than the male body. Consider this editorial from the 1880s which supported the idea of women practising medicine as well as dissection and surgery. The writer initially asked the fairly common question as to why 'woman' was so admirable as a Sister of Charity or a Florence Nightingale and 'so much less angelic as a doctor?'. The truth is, he suggested, such women encountered far more revolting and unpleasant circumstances than most male doctors ever do, and that women in fact have 'the stronger stomach, morally and physically' by virtue of their domestic duties. Then he continues:

> [A]nd as for [women's] impatience of assaults upon their ethical sense, how many men could acquiesce so easily in the frightfully gross and brutalising scenes of a prostitute's life? That women are capable of such extremes of conduct, that they can be so refined and yet so coarse, so very nice and yet ... so very filthy, is a puzzling paradox enough ... but it is a paradox that must be taken into consideration if we want to account for her readiness, and even anxiety to face the horrors of the operating table and the pathologists' shelves.

In this account, the common enough slide between woman as 'Sister of Charity' and 'prostitute' was seen to be characteristic of those few women seeking entry to the university. This construction was brought to bear not only on abstract 'woman' but on particular women. As this writer saw it, the filthy/refined nature of 'woman' meant that actual women were entering the 'unclean profession of medicine' by no means unprepared or unsuited.[29] Of course, this was something of a reversal of the more usual argument that women's purity and delicate sensibility would be defiled and contaminated in the dissection room. However, it suggests also the sense in which women's purity/pollution were so discursively connected, were so much a part of each other, that they could be readily reversed. It was argued sometimes, then, that there was a sexual/moral/physical dirtiness about 'woman' which should be taken into consideration in the debate on particular women who wanted to be doctors – that this dirtiness prepared women for the fleshy, filthy business of dissection.

For women to wield the surgical knife was to have them take up another position intensely loaded as masculine. Recently Rosemary Pringle and Susan Collings have pondered the cultural taboos around women incising flesh of a different kind. Their work on women and butchery is more than relevant in this context. They show how, historically, the practice of cutting that particular kind of meat has also been sexualised, and how clearly the knife stands in for the phallus.[30] In the context of the gendering of nineteenth-century medical practice too, the knife is too phallic to contemplate its use, its appropriation by women: knives belong to men. In an essay in support of women doctors published in 1864 Emily Faithfull dealt at length with the type of medicine women should and could practice. Surgery and dissection were specifically excluded: 'God forbid,' she wrote 'we should ever see the knife in a woman's hand; but why object to her dressing and healing the wound the knife has made?'[31] Similarly, Dr Charles Drysedale argued against women's use of the knife in his paper 'Medicine as a Profession for Women' thus:

> There are four principal divisions of the art [of medicine] made, viz., Physic, Surgery, Obstetrics and Hygiene. Probably of these four constituent parts of the art we may presume ... that women will prefer the divisions where less use of the knife is required than it is in surgery.[32]

The whole question of knives, flesh and their cultural connection with destruction very much shaped the gendering of dissection and surgery. The idea of dissection was not that removed from the idea of killing, the dissection room a place where death and knives were brought together, not quite for the purpose of murder, but not far off it. Moreover, the popular memory of violent and illicit behaviours around corpses and body-snatching was far from repressed.

If the concept of women with knives was too close to the possibility of women destroying and killing, the specifics of this anxiety in men was surely the threat of castration. Here are the concerns of the prominent surgeon and gynaecologist Spencer Wells, as he imagined the nightmare of women as surgeons and men as patients:

> one of them, sitting in her consultation chair, with her little stove by her side and her irons all hot, searing every man as he passed before her; another gravely promising to bring on the millennium by snuffing out the reproductive powers of all fools, lunatics, and criminals; a third getting up and declaring that she found, at least, seven or eight of every ten men in her wards with some condition of his appendages which would prove incurable without surgical treatment, and a bevy of younger disciples crowding around the confabulatory table with oblations of soup-bowls of the said appendages.[33]

Fear of castration haunts this debate on gender, knives, dissection and, as Pringle and Collings note, woman herself embodies such a fear.[34]

Also informing, shaping and limiting the possible meanings and practices of the woman-dissector was a powerful late nineteenth-century feminine/feminist discourse working around the question of the integrity of the body, an argument about maintaining the wholeness of the body. This was a discourse which, at a stretch, might be thought of as a politicisation of bodily boundaries. To start with, the problematic of women and dissection was certainly part of the larger questioning of vivisection, the scientific experimentation on live animals. While some prominent women doctors, in particular Elizabeth Garrett Anderson in England, argued in favour of such experimentation, the anti-vivisectionist position was linked with that particular set of feminist ethics discussed in the last chapter. The anti-vivisection movement was itself directly related to feminist charges of, and challenges to, medical sexual violence against women. Anti-vivisectionist literature of a more feminist bent energetically mobilised the metaphor of medical science as rape, and vivisection as a form of sadistic sexuality.[35] It is important to note here the place of male medical students disrupting and heckling at anti-vivisection meetings.[36] On another tack, Frances Power Cobbe speculated over and consciously manipulated the possibility of 'Jack the Ripper' being a doctor – a physiologist.[37] Well within this same discursive field, the medical use/abuse of the speculum was then being thrown into public consideration, particularly by Josephine Butler in her feminist objections to the Contagious Diseases Acts. Butler argued in terms of the 'sanctity of [a woman's] person' which use of the speculum violated.[38]

One of the axes around which each of these debates was formulated was the problematic of inside/outside the body. Some women themselves circulated powerful discourses which constructed invasion of the body through dissection and surgery, amongst other practices, as masculine, violent and abusive. One gendered and politicised boundary in the domain of health practice was the surface of the body – the skin. In the specific question of dissection, men were culturally permitted to break the surface of the skin, to incise flesh, to look inside the body and dismember it. Women in the medical domain were encouraged to stay outside the body, and to deal with it as a whole, whether that body be dead or alive. And it is clearly the case, as argued in the previous chapter, that many women doctors positioned themselves and their work within a discourse of hygiene and preventive health which both permitted their medical practice to be inflected morally, and distanced them from practices like dissection and surgery, which broke these gendered and politicised bodily boundaries.

TROUBLING BOUNDARIES: FEMALE BODIES, DEATH AND ABJECTION

There was great cultural ambivalence about the dead body which needs to be considered in the debate about women and dissection. The corpse was subject to telling extremes of representation. As Bronfen and others have suggested, there was an imaginative 'cult of the beautiful dead' and yet the dead body was also a spectacle of some horror. The trips to the morgue in Zola's *Therese Raquin*, for example, operate through both revulsion and fascination. 'Morgues were visited like picture galleries, wax museums,' writes Bronfen, in ways which conflated the 'fascination of the preserved dead body with aesthetic pleasure.'[39] In the domain of dissection, as I have suggested, there was a profound disjunction between the imaging of young, sexual female corpses, women who had just 'fallen' from their state of purity, and the male bodies actually dissected, which were marked in various ways with meanings of poverty, destitution, and criminality, and which in hospitals were seen to be the source of contagion and disease.[40] The corpse fitted a sanitary discourse: the body to be dissected was both sexualised and horrible, sanctified and polluted. This instability was compounded by the fluid and changing understandings of the dead body; the open question over whether the corpse still housed the soul or whether it was inanimate flesh – a carcass.

Such ambiguity and ambivalence about the dead body should not be surprising, for death was an unsettled and unsettling site. Questions of dissection aside, simply contemplating the dead body took place in an unstable, marginal domain, for the dead body, as Bronfen writes, is 'neither entirely in the Beyond, nor entirely Here'.[41] Dissection rendered death even more unsettling, in that it interfered with the normal process of transition; it suspended and worked in precisely that moment between Here and the Beyond, between life and proper death marked by burial. Dissection happened, then, on decidedly precarious cultural boundaries, not only between life and death, but also between inside and outside the body, between wholeness of the dead body and its decay, between the material and the spiritual. Moreover, the dissector occupied a somewhat sinister, at the very least unsure, cultural ground between aggression and altruism/healing. He learned about the human body, but through the destruction of that body. As Foucault noted, with the emergence of pathological anatomy, 'knowledge of life finds its origin in the destruction of life and in its extreme opposite; it is at death that disease and life speak their truth'.[42] Given Victorian culture's concern with ordering boundaries of all kinds, the practice of dissection was supremely challenging.

Mary Douglas wrote of men in marginal, borderline positions being 'licensed to waylay, steal, rape. This behaviour is even enjoined on them. To behave anti-socially is the proper expression of their marginal condition'.[43] Her general observations cast in a new light specific stories about the dissection room as a 'sports ground' or as a 'rowdy place; a sort of common room for idle students'.[44] In Douglas's scheme, the corpse (especially when dissected and in pieces) can be seen as 'matter out of place,' matter which cannot quite be stabilised. It represents disorder, dirt, pollution, danger. Following Douglas, Julia Kristeva has developed elaborate ways of thinking about the social and cultural efforts to stabilise borders, especially the borders which constitute the social self. Both culturally and individually, she argues, proper boundaries are defined through the process of expulsion of the impure: abjection. The impure, the abject, can never be completely expelled or removed, but constantly threatens to recur, to return, to pollute. Importantly for this discussion of dissection, Kristeva writes that what causes abjection is 'what disturbs identity, system, order. What does not respect borders, positions, rules. The in-between, the ambiguous, the composite...abjection is above all ambiguity.'[45] If Kristeva gags, wretches, spasms over the skin on the surface of milk, the image/sensation with which she opens *Powers of Horror*, such expressions of disgust and horror are heard over and again in historical texts on dissection, as individuals, men and women, encountered cadavers for the first time. Maria Montessori wrote: 'I felt an extreme weakness and then a feeling of anxiety as if little by little my body were dying. I was leaning against that wall, beside myself, suffering tortures...a horrible smell'.[46] In Kristeva's scheme, the corpse is a specific category of abjection, of bodily waste. Moreover, she identifies the corpse as the most extreme of abjection:

> refuse and corpses *show me* what I permanently thrust aside in order to live...There, I am at the border of my condition as a living being. My body extricates itself, as being alive, from that border.[47]

As Elizabeth Grosz summarises Kristeva, the corpse 'is intolerable because, in representing the very border between life and death, it shifts this limit or boundary into the heart of life itself.... The cadaver poses a danger to the ego in questioning its solidity, stability and self-certainty.'[48] Montessori wrote, as if validating Kristeva's idea that the abject, in this case the corpse, is neither subject nor object: 'I felt as if a very thin thread were connecting my flesh to it.'[49] Further, the boundaries blurred and destabilised in the dissection room are precisely those crucial boundaries at work in Kristeva's concept of the abject: the border between inside and outside. If the abject in Kristeva's terms, or 'marginal stuff' in Douglas's terms, is that which issues from the inside of the body out, the dissector reverses/transgresses this process of expulsion. Rather than pushing out and away that which is unclean, the dissector breaks through the most fundamental social boundary of the body – the skin – revealing, handling, immersing in the inside of the already-abject body-corpse.

The conceptual processes of abjection and dissection work through similar insecurities. In the same way that abjection is about casting out that which cannot be fully cast out, in dissection rational medical science attempted to know fully that which is ultimately unknowable – death. In what might function as the perfect metaphor for the positivist scientific dream of the reified object, the dissector attempted to stabilise and know the profoundly unstable borders represented by the corpse; borders between life and death. This is a state which culturally and scientifically demands to be fixed, but cannot be. The corpse is abject in the sense of representing human waste, but also as it refuses to be a knowable, reified object.

The dissection of the specifically female corpse brought together what Bronfen identifies as the two central enigmas of Western culture:

> Death and femininity are culturally positioned as the two central enigmas of western discourse. They are used to represent that which is inexpressible, inscrutable, unmanageable, horrible; that which cannot be faced directly but must be controlled by virtue of social laws and art.[50]

Bronfen writes that '[d]eath and femininity both involve the uncanny return of the repressed... [they] cause a disorder to stability'.[51] The female corpse combined two elements disruptive of order, signifying both the attempt and the failure of 'man' or rationality or science to expulse the Other, 'a superlative figure for the inevitable return of the repressed'.[52] Science, desiring to know, fix, stabilise and explain most everything, was the ultimate tool to attempt to control the enigmas of death and femininity. The imaginative and symbolic feminising of the corpse can be seen as an attempt to render it object, not abject, to bring to bear on the corpse the gendered medical and scientific discourses which inform who and what is subject and mind, who and what is object and body.

Dissection was a site where the resistance to female medical practice was most energetic, and where the gendered and embodied subject position of the practitioner was clearly at issue. There existed a cultural if not quite literal boundary in the domain of medicine which was strongly gendered – inside of bodies for men, outside of bodies for women. But this was not simply an imposed boundary, as far as women doctors were concerned. It was at least in part created and authorised through discourses circulated by themselves and their feminist proponents. Women as well as men had difficulty contemplating female medical students observing, incising, and handling the dead body. On one level, this uneasiness and sometimes outright physical protest, needs thinking through in a strictly historicised way: in terms of the specificities of late nineteenth-century systems of gender and sexuality; in terms of the discourses of resistance to these systems; and in terms of the unstable meanings of both death and the corpse in a moment of Western culture which, while still deeply religious, was nonetheless rapidly secularising and scientising. On another level, this unease about dissection can be read via concepts which are not strictly historical. In particular the theoretical problematisation of mind and body, subject and object, and the feminist recognition of the gendering of these dualisms, very clearly operated in the late nineteenth-century dissection room. And if obsession with order and clear boundaries characterised dominant Victorian culture, theories which deal precisely with the imperative toward, and the impossibility of, such order and clear boundaries can enrich conventional historical readings.

Within medical history, Jordanova's work paved the way for critical examinations of sexual/medical encounters between the feminised object–body and the masculinised subject–dissector/surgeon. Jordanova argues that '[i]t matters not the slightest what the biological sex actually is of the body under the knife, because sex here is the perfect metaphor for a particular admixture of power and pleasure'.[53] Yet I argue that the actual sex of the corpse is more significant than Jordanova allows. Both in terms of the efforts of male dissector to disavow the objectification and penetration of other male bodies, and in terms of the difficulty women had with male corpses specifically, the problematic encounter between dissector and corpse always involved two sexed bodies, not one. The presence of the female medical student suddenly foregrounded sexual difference. This rendered the male student embodied and sexed and undermined the possibility of their fantasy of disembodiment. Moreover, the figure of the female dissector intensified the signification of disorder, pollution, the repressed already superlatively present in the act of dissecting the corpse.

7 Sterile Bodies: Germs and the Gendered Practitioner

If any topic is privileged in the history of nineteenth-century medicine it is the germ theory and the supposed revolution in surgical practice which was its result. Many historians write of sanitarian and miasmatic conceptualisations of disease and health-care practice as being replaced around the turn of the century by the truths of a body of medical knowledge produced by the scientific method rather than by an earlier empiricism. In A. J. Youngson's story of progress, *The Scientific Revolution in Victorian Medicine,* for example, miasmatists simply misunderstood disease or did not understand it at all: 'They showed good sense; although it cannot be said that they possessed good science.'[1] In such accounts, a pre-existing 'reality' of germs and all that went with them is constructed as being uncovered, or discovered, by an increasingly 'true' medical science. When one turns to critical studies of thought on germs, dirt and pollution, studies which suggest that science and culture actively produce rather than simply reveal the object of study, or 'nature', the paradigm-shifting revolution based on knowledge of germs is still there. Mary Douglas writes:

> [O]ur idea of dirt is dominated by the knowledge of pathogenic organisms. The bacterial transmission of disease was a great nineteenth-century discovery. It produced the most radical revolution in the history of medicine. So much has it transformed our lives that it is difficult to think of dirt except in the context of pathogenicity.[2]

On any reading, not least in the reading of contemporary medical texts, large changes did take place at the end of the nineteenth century. There was certainly a broad shift between 1860s sanitarian ideas underpinned by miasmatic theories and turn-of-the-century bacteriological theories of disease. However, in the unrelenting focus on this period as witnessing a paradigm shift, a 'revolution' or a major transitional generation of medical thought and practice, there is a marked failure to note and explain the ongoing relevance and currency of supposedly outdated sanitarian ideas into the twentieth century. This continuing viability of 'older' ways of thinking about health, ill-health, dirt and disease demands analysis.

What I propose in this chapter is not to challenge the idea that a major shift had taken place by the turn of the century so much as to reinterpret the shift as far more complex than is usually suggested. There was no straightforward adoption of a self-evidently truthful 'germ theory' which rendered the sanitarian world view meaningless or impossible to contemplate. The emergence of germ theories of disease demands a more complicated and nuanced reading than the simple invocation of a paradigm shift or a 'scientific revolution'. I argue that sanitarian concepts of healthiness and cleanliness, as well as specific 'older' forms of medical therapeutics, were not so much replaced in this period as displaced largely into a female domain of work. Sanitarian discourse of health and disease was sustained in the field of nursing knowledge and practice well into the twentieth century. Moreover, the technologies of cleanliness produced by ideas of 'asepsis' allowed for, rather than obliterated, nurses' sanitarian practices. I also suggest that while 'germ theories' produced radically different practices and ideas about dirtiness/cleanliness for doctors, there was no identical implication for nurses. While the turn-of-the-century concepts of asepsis asked doctors to think of themselves as potential polluters, thereby needing to sterilise themselves, nurses had thought of their work and themselves within a language of absolute purity for a generation and more.

ANTISEPSIS AND ASEPSIS

Antisepsis, as it was developed and practised in Britain in the late 1860s, 1870s and 1880s, was based on the principle of excluding microbes from open wounds, killing them with some sort of antiseptic medium. Lister and some other surgeons developed complex and ever-changing methods of wound dressing, creating barriers of various sorts – gauzes, bandages, packing, putty – between the wound and the open air. The barrier was to be soaked or in some way permeated with a disinfectant, usually carbolic, which supposedly destroyed any microbes nearing the wound. It is important to note, however, that the use of carbolic itself was far from an innovation on the part of Lister.[3] A later development was the famous carbolic spray, the machine which infused carbolic acid into the atmosphere surrounding the open wound during the operation, with the intention of destroying the airborne microbes. Like existing miasmatic ideas, antisepsis continued to problematise the air as the main medium of infection, in a way which permitted many surgeons to adopt particular practices without radically changing their concepts of disease causation.[4] Lister himself continued to structure his theories and principles around the idea of germs in the air rather than on hands, clothing or instruments.

From about 1890, methods of disinfection began to turn from an emphasis on antiseptic chemicals to the aseptic practice of sterilisation by heat and steam, which had been developed mainly in Germany.[5] While there continued to be a certain amount of ambiguity and confusion about the usage of the terms antisepsis and asepsis, indeed they were often used interchangeably, some surgeons and pathologists began to make a distinction of principle between them. Charles McBurney, for example, wrote: 'By "Antisepsis" is meant the adoption of various methods of destroying bacteria, or inhibiting their growth... asepsis means absence of germs which produce sepsis.'[6] And elsewhere: '[T]he term *aseptic* conveys the idea of freedom from all forms of bacteria, putrefactive or otherwise; and the term antiseptic is used to denote a power of counteracting bacteria and their products.'[7] Thus, for example, instead of rendering a wound germ-free by soaking a dressing in carbolic or some other antiseptic, the dressing itself was sterilised, rendered germ-free, and then applied to the wound. Gradually, more and more items, ideally everything which came into contact with the surgically opened body, came to be sterilised: scalpels, retractors, forceps and other instruments, sponges, dressings, silk and other materials used for suturing, towels, drainage tubes.[8] By 1910 there was a reasonably wide medical acceptance of the principles of asepsis, and of the possibility of achieving a completely sterile, germ-free environment in which to undertake surgery or to change dressings on wounds. Most significantly, the possibility of asepsis turned attention to the hands, fingers, arms and face of the surgeon. I shall return to this crucial point below.

The explicit confusion about disease aetiology and wound infection which characterised mid nineteenth-century debates had become a markedly uniform (though by no means universally accepted) body of knowledge about germs. In the last decades of the century, this knowledge was institutionalised as the new disciplines and specialties of bacteriology and microbiology.[9] In one sense contagionist theories had been validated. However, a major difference between contagionism of the 1860s and turn-of-the-century germ theory was the discovery that the microbe in question was not 'morbid matter', but a living organism which reproduced itself rapidly. Surgery and bacteriology were thoroughly linked by the turn of the century. Bacteriological knowledge was coming to be seen as the grounding knowledge for surgery, replacing, for example, the manual, visual, tactile and experiential knowledge of dissection. The first sentence in Da Costa's 1900 text, *Modern Surgery*, was not about surgery at all, but about bacteriology, 'the science of micro-organisms'.[10] This opening chapter on bacteriology presents itself as the fundamental surgical knowledge, a knowledge which 'explains' the rest of the book's content on wounds, procedures, precautions, antisepsis and asepsis. As stated in another textbook from 1906: 'So much has bacteriology come to dominate every department of surgery in particular, that it is now the standpoint from which nearly all surgical questions have to be considered.'[11]

It is important to note, however, that despite this large change in thinking about wound infection, miasmatic theories retained a certain currency. There was no simple succession of paradigms, nor a straightforward shift from the idea of multiple causes to the aetiological concept of single and necessary causes. The explicit endorsement of the theory of direct contagion did not necessarily require doctors to abandon concern for environmental or miasmatic causes. An 1898 edition of Playfair's midwifery text is a perfect example of two apparently contradictory aetiologies appearing quite literally side by side. At one page in this edition a plate of streptococci is illustrated, accompanying a very detailed account of the action of pathogenic organisms. This did not stop the author from detailing a case study from his own practice which pointed directly to the defective sanitary state of the house as the cause of puerperal fever. So accompanying the plate of streptococci is another illustration, an architectural one, which displayed the architectural positioning of the guilty concealed water-closet: the 'source' of the disease had been revealed.[12]

Received historical stories about antisepsis, asepsis and the history of surgery are being revised. In a standard historical account, antiseptic practice in the 1870s and 1880s is seen to have been strongly resisted. Gradually, however, commonsense prevailed, the 'truth' of germ theory became unarguable, and antiseptic practice was finally accepted by surgeons. More recent medical histories make such a straightforward story untenable. Significantly it was possible, if not common, for doctors to adopt the use of carbolic or another disinfectant, to use a spray during their operation, or to alter their procedures for dressing surgical wounds, yet at the same time to ignore or deny any fundamentally different principle of infection. Antiseptic techniques did not necessarily demand any conceptual shift in thinking about disease and illness. Many surgeons came to utilise antisepsis simply as an alternative surgical technique, for which there were appropriate and inappropriate cases.[13] Others developed an empirically-based faith in carbolic to prevent those infections which plagued hospitals and surgeons' statistics, again without any necessary subscription to the idea of germs. Christopher Lawrence and Richard Dixey have suggested that a major inconsistency in the received story about Listerism is the fact that there was not one but many competing theories on germs and wound putrefaction in the late nineteenth century. These authors have convincingly argued that Lister's early (1860s) germ theory of putrefaction (after Pasteur) was quite different to, if not inconsistent with, his late nineteenth-century germ theory of infection (after Koch). Lister and his supporters, they suggest, successfully remoulded their initial antiseptic practices and their research to fit subsequent theoretical developments in microbiology and bacteriology. Antisepsis is often understood to have been the immediate theoretical and practical precursor to modern aseptic techniques of surgery. Even at the turn of the century, there was already a standard account that the new aseptic techniques were a natural development of Listerian antisepsis.[14] But according to Lawrence and Dixey, Listerian practice came to be seen as the revolutionary precursor to asepsis not by virtue of an actual link, but because advocates managed to connect it successfully and in a sense retrospectively with the idea of 'the' germ theory: 'Listerians could represent the practice of aseptic surgery, which was based on the by now recognisably modern germ theory of infection, as deriving from a simple elaboration of early Listerian ideas and practice.'[15] It is suggested by them, and by several other medical historians, that aseptic surgery actually emerged from non-Listerian traditions.[16]

The connection these historians make between asepsis and public health/sanitary reform, rather than between asepsis and antisepsis, is very interesting and illuminates far more clearly the concerns, technologies and concepts of turn-of-the-century hygiene, social and surgical. Yet, insofar as their arguments, research and frameworks are uninformed by the problematic of gender, whole areas of continuity and discontinuity around the question of antisepsis, asepsis, surgical hygiene and public hygiene are obscured, and misinterpretations are permitted. Different practitioners of these theories and technologies – male and female nurses and doctors – need to be understood as embodied and as gendered, and therefore as having differing investments in, and relations to, discursive practices involving dirtiness and cleanliness.

NURSING KNOWLEDGES AND PRACTICES

In the history of nineteenth-century nursing there is a widespread subscription to the importance of the shift between theories of disease. Sandra Holton's article on Florence Nightingale, for example, takes the shift from miasmatic/sanitarian ideas to germ theory/scientific ideas as given, and as central to understanding changes in this domain. She writes of the implications of the new reductionist and scientific discourse of health-care thus:

> When disease phenomena were reduced to abnormalities in particular organs or functions, rather than the disturbance of a whole body system, disease agency was now located in microscopic organisms. The triumph of this corpus of thought demanded a kind of technician, a technology and a location for health care which were entirely different from those proposed by Florence Nightingale. The technical competence required of nurses was increasingly dominated by the requirements of this new 'scientific medicine' and Florence Nightingale's case for women's autonomy in health care, based upon moral authority, was therefore undermined.[17]

As I stated in the introduction to this chapter, I do not want to dispute that there were major changes over this period. But I do want to question and qualify the revolutionary 'success' of germ theories, of bacteriology, of reductionist scientific medicine; that is, I want to question whether there *was* a clear 'triumph of this corpus of knowledge'. Close examination of nursing texts published into the early twentieth century suggest not a radical shift of paradigms but a remarkable continuity. Certainly, reductionist models of the body enabled by the new microscopic vision, by bacteriology and by a distinctly modern sensibility which privileged the breakdown of knowledge to its smallest units, can be identified in these texts. However sanitarian concepts remained entirely viable and arguably dominant throughout the period. 'The germ theory' might not be so revolutionary or paradigm-shifting as many histories suggest. If asepsis was based rather more on sanitarian and public health discourse than on established antiseptic practices, then this permitted a continuity not a disruption of dominant ways of thinking about nursing. The 'purity' aspired to in asepsis can be seen to be a version of that scrupulous cleanliness so central to mid nineteenth-century sanitarian nursing. Before I examine antisepsis and asepsis specifically, it is worth turning to the larger context of nursing knowledge and practice within which the 'germ theory' and accompanying practices were assessed.

Many nursing texts written by women continued to focus on a set of practical skills which remained remarkably constant from the mid nineteenth century well into the twentieth century: the application of leeches, cupping, forming blisters and the preparation of remedies such as lotions, poultices, fomentations, baths, enemas.[18] In Chapter 1, the ongoing currency of miasmatic/sanitarian concepts was discussed in terms of women's capacity to locate themselves with authority within this discourse, and to exert some control over it. There is a further aspect to this. It is evident from nursing texts that many specific older therapeutic interventions based in humoral theories about balancing the body were sustained in, perhaps transferred to, this female domain of health-care. This should not be viewed simply as the preservation of an outdated paradigm as doctors moved forward toward some sort of truer science, but as a crucial perpetuation of ideas about health, disease and the body which never really disappeared, as advocates of a nineteenth-century 'scientific revolution' would have it. Earlier models and practices were not overtaken, but were largely incorporated into and often conventionalised and ritualised in, the work done by women as nurses in hospitals and in homes. Much of what was considered masculine medical knowledge and practice earlier in the nineteenth century became feminine knowledge and practice by the turn of the century. Yet in this process of feminisation, such practices came to be radically devalued, losing their status as scientific knowledge.

In an illuminating and important article on medical practices, Charles Rosenberg has examined the traditional therapeutics of early and mid nineteenth-century America, and the complex configuration of ideas which resulted in a very different therapeutic system by the end of the century.[19] In the traditional system, the body was seen in constant interaction with its environment. Health or disease were not locally produced states, but 'general states of the total organism'; balance and equilibrium meant health. The physician's most potent weapon, writes Rosenberg, 'was his ability to "regulate the secretions" – to extract blood, to promote the perspiration, urination, or defecation' primarily with the use of drugs. Effective therapies relied on physiological effects which were visually or sensually perceptible by doctor and patient. It was also crucial that the physician create 'an emotionally as well as intellectually meaningful therapeutic regimen.'[20]

Much of what Rosenberg presents as early nineteenth-century therapeutics resonates in nursing texts of the late nineteenth and early twentieth centuries. At the most basic level, many specific therapies and interventions devolved onto nurses. Forming blisters and other methods of counter-irritation, for example, were part of the traditional physician's methods. The purposeful causing of suppuration on one part of the skin would, as Rosenberg explained, attract 'the morbid excitement from another site to the newly excoriated one, while the exudate was significant in possibly allowing the body an opportunity to rid itself of morbid matter, of righting the disease-producing internal imbalance'.[21] Eva Lückes's widely used text included a whole chapter on counter-irritation by forming blisters, by 'cupping' or by the application of hot poultices. She explained it as the formation of an irritant on the skin, an inflammation, which counteracts a 'deeper-seated' inflammation'.[22] Extremely detailed instruction on such methods remained a stock part of nursing knowledge into the twentieth century. In twentieth-century texts written by nurses, whole lectures were still devoted to counter-irritation, which retained a clear sense of correcting an internal bodily imbalance. A 1935 Australian text, for example, taught that '[e]ach organ is reflexly related to some area on the outside of the body, so that counter-irritants applied to those areas will also affect the associated deep-seated organ'.[23]

Traditional therapeutics also relied on a careful and trained use of the senses, of smelling and touching particularly.[24] By the end of the nineteenth century, new medical technologies were either mediating such practices for doctors, or rendering them outmoded. This use of the senses as a form of diagnosis and observation was claimed by nurses: 'Hospitals are especially places for the observation of disease. It is in these institutions that medical knowledge is made, so to speak...Nurses should use their senses in due order, and mentally record what these senses teach them. Sight, touch, smell, hearing, have all to be trained to do their duty...Certain diseases have characteristic odours which experience alone can render familiar to you.'[25] In this account, technologies were eschewed: 'Doctors aid these senses with the microscope, the clinical thermometer, the stethoscope, and many other appliances; but, of course, nurses can ascertain all that it comes within their province to know, without the aid of all these things.'[26] Just when medical practice came to involve the use of such technologies, nursing practice consolidated and perpetuated older forms of observation and diagnosis.

The idea of the sick body requiring a balancing and regulating therapeutic intervention also retained a currency in nursing practice. The earliest nursing texts in the 1860s and 1870s were formulated by a generation of medical men educated in the 1840s and 1850s, for whom humoral concepts of the body and of therapeutics were barely under challenge. Foundational texts such as Munro's 1873 *Science and Art of Nursing the Sick,* wrote into several generations of nursing knowledge the importance of carefully observing and regulating a set of bodily fluids: 'The nurse should acquire at first a *definite, accurate,* and *concise* knowledge of what the healthy *secretions* and *excretions* are, and fix it in her mind for all time to come.'[27] A range of practices ensued which, although reworked and ritualised in various ways, remained meaningful well into the twentieth century. The constant monitoring, measuring, recording and regulating of input and output, of eating, vomiting and defecation, of drinking, urinating and sweating came to be central to nurses' understandings of their role. Moreover, instructions to manage hospital wards in particular ways also drew on this deeply embedded notion of the need for balance. To extend the analogy between rooms/houses/wards and bodies developed in Chapter 1, an analogy which was always more than metaphorical, it was quite common in nursing texts for wards, like bodies, to require an imposed regime of regularity. 'The sister is responsible for the entire well-being of her ward,' wrote Catherine Wood, 'for its comfort, order, cleanliness and regularity.'[28]

The notion of rest, of responsibility for creating and maintaining a calm context for the sick person, was also insisted upon. A 1903 text instructed: 'The general neatness of arrangement in the sick-room is essential, recovery being assisted by pleasant surroundings, soothing the mind and acting as a sedative upon the irritated, over-strained nervous system.'[29] The notion was perpetuated in the nursing domain that it was not only specific medical, surgical, pharmaceutical interventions, which led to recovery, but an environment cared for and managed in a particular way: a room which was 'nursed', an atmosphere which was purified; a room that was calmed, a family that was reassured. What came to fall within the nursing domain of treatment, then, were remedies and therapies which visibly altered and regulated the body, as well as sanitarian concepts in which the environment was part of health and disease: the application of heat and cold; of perceptible irritants; of electricity; the regulation of light, rest, sunshine, fresh air. The nursing of people sick with diphtheria is a case in point.

> The human skin is admirably and obviously constructed for the precise purpose of ridding the system of its waste matter... Wrap the child closely in a blanket well rung out of hot water, with plenty of dry blankets over it, keep it in these from twenty to thirty minutes then sponge quickly with tepid water (in a warm room)... with a hot water bottle to the feet... *It opens the pores,* sets them working, equalises general while reducing local heat, and relieves obstruction everywhere... Help nature by making the skin moist and soft, and she will take instant advantage of your labour... *Nature is ever ready to cure* she is remedial in her efforts. It is wonderful what power she possesses of putting things right.[30]

In this paradigm, imagining the body in its smallest physiological unit, the cell or the germ, was relatively meaningless.

New microscopic and surgical sciences and practices gradually replaced earlier therapeutics in the medical domain. In most historical accounts, this is seen to be the end of the story. But if late nineteenth-century medical and nursing practices are assessed as *jointly* constituting the domain of health (a remarkably uncommon historical assessment), then not only were the sanitarian concepts explored in Chapter 1 quite viable in the early twentieth century, but many old heroic therapies which worked through an impact visibly and sensually apparent to both patient and practitioner had been sustained as well. What facilitated, or was even a precondition of the development of the new medical therapeutics, was the effective displacement and incorporation of the 'deeply internalized'[31] emotional/moral/environmental conceptualisations of the sick body into the nursing/female domain of practice.

How then did the specific knowledges and practices informed by medical theories about germs, antisepsis and asepsis fit into this well-established domain of nursing practices? Overall, for nurses, both 'antisepsis' and 'asepsis' seem to have functioned as a new way of coding 'cleanliness'. In many instances, far from undercutting the broader sanitarian concept of cleanliness, antisepsis and especially asepsis were taken as congruous. Thus, while Lister himself might have suggested that hospital cleanliness was relatively unimportant so long as antiseptic methods were strictly adhered to in surgery, established sanitarian ideas were simply too dominant to be so easily displaced. Indeed, in nursing texts examined up to 1910, germs, antisepsis and later asepsis were by no means central organising principles.

Contrary to Holton's position it can be argued that in nursing discourse, antiseptic procedures and accompanying germ theories, whether fully accepted or not, did little to disrupt established practices of and ways of thinking about sanitary cleanliness and hygiene. When the language of germs *was* employed, as in Eva Lückes *Lectures on General Nursing*, this was not accompanied by, nor did it demand, any radically different practice in hospitals. Lückes instructed women that '[t]he generally accepted theory is that infecting germs may be dispersed in a variety of ways – wafted by the air, carried by water and milk, or conveyed by our clothes'. Practices stemming from such ideas, even those which incorporated a notion of antisepsis, sustained the dominant concept of 'foul air', which was necessary to make sense of so many specific nursing interventions. For example, in caring for fever patients it was suggested that a sheet soaked in disinfectant be hung in a room or over a door: 'If the air is impregnated with poisonous germs, it is a very reasonable theory to load the atmosphere as far as possible with the antidote to that poison.'[32] This was a literal translation into nursing practice of the Listerian idea of carbolic spray functioning as an antiseptic shield or curtain, to prevent the access of atmospheric germs to a wound.[33] Far from heralding or requiring a new conceptualisation of disease, it was perceived as simply a new technique of purifying the air. For nurses, the air remained the problematised medium.

Catherine Wood's text included detailed descriptions of complicated antiseptic dressings involving carbolic spray. But it was not organisms *per se* which she isolated as the problem, but the atmosphere: 'All dressings which aim at the prevention of the formation of pus, by excluding the air and by purifying the surrounding atmosphere and the surface of the wound, are *antiseptic... the pus must always be excluded from the air.*'[34] For both of these writers, fresh air and sunshine were still fundamental therapies which the notion of antisepsis had done little to overturn. 'Dark corners are not healthy,' wrote Lückes. 'You may notice that any dark or shady corner will retain a disagreeable smell... Sunshine is a necessity physically as well as morally, and it has a definite and powerful influence for good in many ways.'[35] Wood rendered her version more scientific, preferring the use of 'oxygen' to fresh air, but neither the concepts of germs nor antisepsis nor even asepsis in this period, shifted the primacy of nurses' mission to purify the air: 'Infection or contagion is an enemy whose existence must be recognised, and with whom doctor and nurse must do battle... He is invisible, only betraying his presence by his deeds... but he has a dire foe in oxygen, and no more able partisan can be called into our aid than this all-pervading health-giving gas.'[36] Sometimes, also, reference to 'the germ theory' was an appeal by nurses to modernity, an attempt to locate themselves within a scientific domain and to incorporate a theory that was gaining cultural status. In such formulations, 'germ theory' was a new way of signifying new nurses' difference from old nurses, and their alliance with medical men, functioning as a code for modernity, progress and superiority.

Nurses' use of carbolic for cleaning (the 'carbolising' of beds for example) as opposed to doctors' use for surgical dressing, simply marked it as the latest in a long tradition of cleaning methods and preparations. It superseded whitewashing, lime, 'dry-scrubbing', all of which at various times were understood to prevent pyaemia, hospitalism, puerperal fever and so on. This conflation of antisepsis with cleanliness meant that antiseptic procedures were often seen to be enhanced by women's particular skill in, and capacity for domestic hygiene. This applied not only to nurses, but occasionally to women doctors. In London, for example, the annual meeting of the New Hospital in 1897, an institution which employed only women, was told that women were

> naturally fitted for hospital control, both by possession of the charitable qualities essential to philanthropic work and by their home training in household economy and management... its medical and surgical staff includes many lady doctors of eminence, who are successfully following the antiseptic treatment initiated by Sir James Paget and Lord Lister, a method for which their own ideas of cleanliness give them especial sympathy.[37]

Asepsis was a concept which crept into nursing texts alongside the explanation of antiseptic practices; it did not cancel the other out. An 1899 textbook by a surgeon typically included both the general principles of antiseptic and aseptic surgery, in a way which prepared nurses for both types of operation still in use.[38] Another wrote of asepsis as a state of cleanliness, as opposed to the mere use of antiseptics as an agent or a technique. That is, while antiseptics might be used as a way of achieving asepsis, the latter is the ultimate condition of cleanliness to which nurses should aspire: 'By "asepsis" is meant the absence of septic germs – i.e. a condition of surgical cleanliness... The great truth that a nurse must always keep before her is, that for the production of asepsis, cleanliness is all important, and that antiseptics play a very secondary part.'[39] This is not to say that nurses were uninformed about developments in microbiological sciences. Herbert Macleod's 1911 text on hygiene did contain very detailed information on 'bacteria', 'bacilli' and 'microbes', and included various practices of disinfection by heat and chemicals which were added to, but did not replace the centrality of purifying the air. Whilst incorporating a specific understanding of infection by germs, the classic sanitarian pattern nonetheless remained intact. His chapters began with 'Air-Pure and Impure', 'Ventilation – Natural and Artificial' and continued with 'Heating and Lighting', 'Water', 'Drainage and Sewerage', 'Personal Hygiene', 'Important Acts of Parliament Relating to Public Health'.[40] This was a continuation of longstanding ways of organising nursing texts which typically began with a chapter not on the patient but on the room, cleanliness and air, entitled 'Ventilation' or 'Hygiene of the Sick-Room'.[41] Macleod's text for nurses suggests the way in which it was possible for microbiology and sanitary reform/public health to be quite complementary and to function well within the same discursive domain. In the context of 'hygiene', there was no necessary mutual exclusion involved in reductionist/germ theory and sanitarian/humoral models. His discussion of germs and infection was primarily about public health, public spaces and the communication of disease through social activity which required modifying: in part through state intervention, in part through public education undertaken by health-practitioners such as nurses. The overt morality which accompanied early sanitary reform was muted, but the regulation of particular bodily behaviours was of more concern than ever: to modify interaction between children, behaviour on omnibuses and trains; to disinfect books from circulating libraries, toys donated to hospitals, the brushes and combs of a hairdresser, and to eradicate 'indiscriminate expectoration'.[42]

Medical histories, even of the more critical variety, usually locate the discovery of germs within the story of wounds, infections and the development of surgery. Yet ideas about germs were not necessarily constituted in the context of hospitals, operations and dressings. Rather, there was a whole field of meaning which brought together the new language of germs and bacteriology with established concepts of hygiene, sanitation, public health, personal cleanliness. Asepsis was by no means a state of cleanliness seen to be exclusively attainable in the operating theatre, as articles such as 'Aseptic Hairdressing' and 'The Asepsis of Towns and Dwellings' attest.[43] Rather, such ways of thinking suggest the strong connection between turn-of-the-century concepts of asepsis, antisepsis and germs and earlier practices of domestic hygiene and personal hygiene in domestic and public spaces. 'The cook and the housekeeper, as well as the surgeon, must study the details of antisepsis', wrote the author of 'Antiseptics in the Larder'.[44] While the new discipline/knowledge of bacteriology became a grounding knowledge for surgical practice, it also intervened in the existing world of 'Hygiene'. Bacteriology was as much about public health and infectious diseases as it was about surgery and wound infection. The author of *Principles of Bacteriology* (1902) for example was a Professor of Hygiene and Bacteriology and director of a 'Laboratory of Hygiene'.[45] Scientific bacteriological and surgical knowledge moved 'outwards' to influence domestic and public notions of cleanliness. Equally, existing notions and practices of domestic hygiene influenced rituals and technologies of cleanliness rendered even more imperative by the invisible presence of germs. A language of purification, well established in terms of nursing and with its roots firmly in a sanitarian sensibility, came to define the cleanliness required in asepsis at the turn of the century.[46] Cleanliness and purity belonged largely to a female domain of work, knowledge and identity. They were concepts already built into ways of thinking about nursing, and about nurses as embodied practitioners.

SURGEONS AND THE TECHNOLOGIES OF ASEPSIS

Nicholas J. Fox's central argument in his article 'Scientific Theory Choice and Social Structure' is that asepsis did not develop from Listerian antisepsis at all, but in opposition to it: that asepsis was 'based *on a completely different theory*' and had far more to do with public health than germ theories. While I clearly agree with Fox's linking of what he calls 'humoral theory' with asepsis, I take issue with his arguments on several grounds. The analytic moves Fox makes in order to equate humoral/sanitarian ideas and aseptic ideas are suspect because he shows no awareness of, or interest in, the fact that 'cleanliness' in the earlier model was primarily about nurses, not doctors or surgeons. It is because of this oversight that Fox is able to see humoral/sanitarian and aseptic models as more or less the same. And so, he fails to attribute any significance to the fairly sudden assumption of rituals of cleanliness by doctors in the aseptic model. It seems to me that it was specifically for *nurses* that aseptic cleanliness was a continuation of earlier technologies, rituals, embodied practices. For doctors and surgeons, however, it was quite new. Doctors were newly implicated in a world of purity, pollution, contamination and scrubbing, a world of 'true cleanliness and scrupulous attention to detail,'[47] which had defined the identity, embodiment and representation of nursing work for several generations. If for nurses the shift from sanitarian ideas and practices through antisepsis and asepsis required no radical overturning of ideas about bodily and spatial cleanliness, for doctors this was not the case.

Fox also suggests that germs were central to theories of antisepsis but barely applicable at all to theories of asepsis: 'With germs excluded, they play no part in the discourse.'[48] On both counts he is mistaken. Firstly it was entirely possible to champion antiseptic practices without subscribing to a theory of germs as such. Secondly, to suggest that 'germs' were not central to the world of surgery, hygiene and public health at the turn of the century, one would have to somehow obliterate not only whole chapters but entire sections on bacteriology from surgical and hygiene textbooks. There are more subtle and historically accurate ways of drawing the connection between sanitarian and aseptic models than by attempting to erase the demonstrable presence of 'germs' in the discourse. By seeing germs as unproblematically present in 'antisepsis' and absent in 'asepsis', Fox is able to suggest that '[a]ntiseptic theory had the effect of equating surgeons with pollution' and that '[a]septic theory equates the surgeon with purity'.[49] His discussion of surgeons' new 'purity', and of asepsis as a 'purifying' process, shows no insight (despite his explicit use of structuralist theory) into the way in which purity gained its meaning through pollution. It seems to me that aseptic rituals and methods were required of the surgeon around the turn of the century, precisely because germs marked him as impure, septic, contaminated. It was *precisely* because surgeons were now implicated as polluters that their new rituals of cleanliness/purity/ sterility became so meaningful and elaborate. That potential to pollute which so many mid and late nineteenth-century doctors had attributed with considerable invective to midwives and old nurses, was now unequivocally inherent in all practitioners. It was now surgeons who were being asked to modify, cleanse, purify, sanitise and sterilise their very bodies. Now they too had to 'scrub'.

Surgeons changed many of their habits and techniques fairly quickly from the late 1890s. As I have indicated, the sterilising of equipment by heat and steam was the first large shift. However the body of the surgeon, now conceptualised as contaminated, clearly could not be sterilised thus. The question of how to render the surgeon's body sterile monopolised much of the literature on surgical procedures. As the difference between asepsis and antisepsis was often seen to be the effective use of heat and steam in the former and the dubious use of chemicals in the latter, the washing of the surgeon was often understood as the one obstruction to 'true' asepsis; everything except the surgeon could be properly and reliably sterilised, and once again hands were the major issue. 'There is little room for doubt that in surgical operations the hands of the operator are by far the most frequent cause of infection.'[50] Far from embodying an unquestioned 'purity' as suggested by Fox, surgeons saw themselves as the stumbling block to perfect asepsis, the item which remained dirty.

In rigorous experiments, bacteriologists and surgeons tested the differential value of soap and water, alcohol, perchloride of mercury, carbolic acid, lysol, formalin, potassium mercuric iodide and ether, in every combination imaginable (see Figure 7.1). They tested various ways in which to scrub the hands and arms: five minutes soaking in one solution, or three minutes scrubbing with such and such an action, with or without a previously sterilised nailbrush, or seven minutes washing with soap and water only. Just one example indicates the minute detail of these procedures:

> a preparatory process has to be gone through, which consists of careful trimming of the nails, washing with ordinary soap and hot water...[then] *Disinfection process*...Washing for five minutes with spirit of green soap and very hot water. The scrubbing is done under a tap of running water, and two sterile nail brushes are used in succession. *Dehydration* by carefully rubbing the hands and forearms with pieces of gauze soaked in methylated spirit; this occupies three minutes. *Disinfection proper*, by rubbing for two minutes with gauze soaked in a 1 in 500 solution of mercuric biniodide in 70 per cent alcohol. The biniodide is washed off with

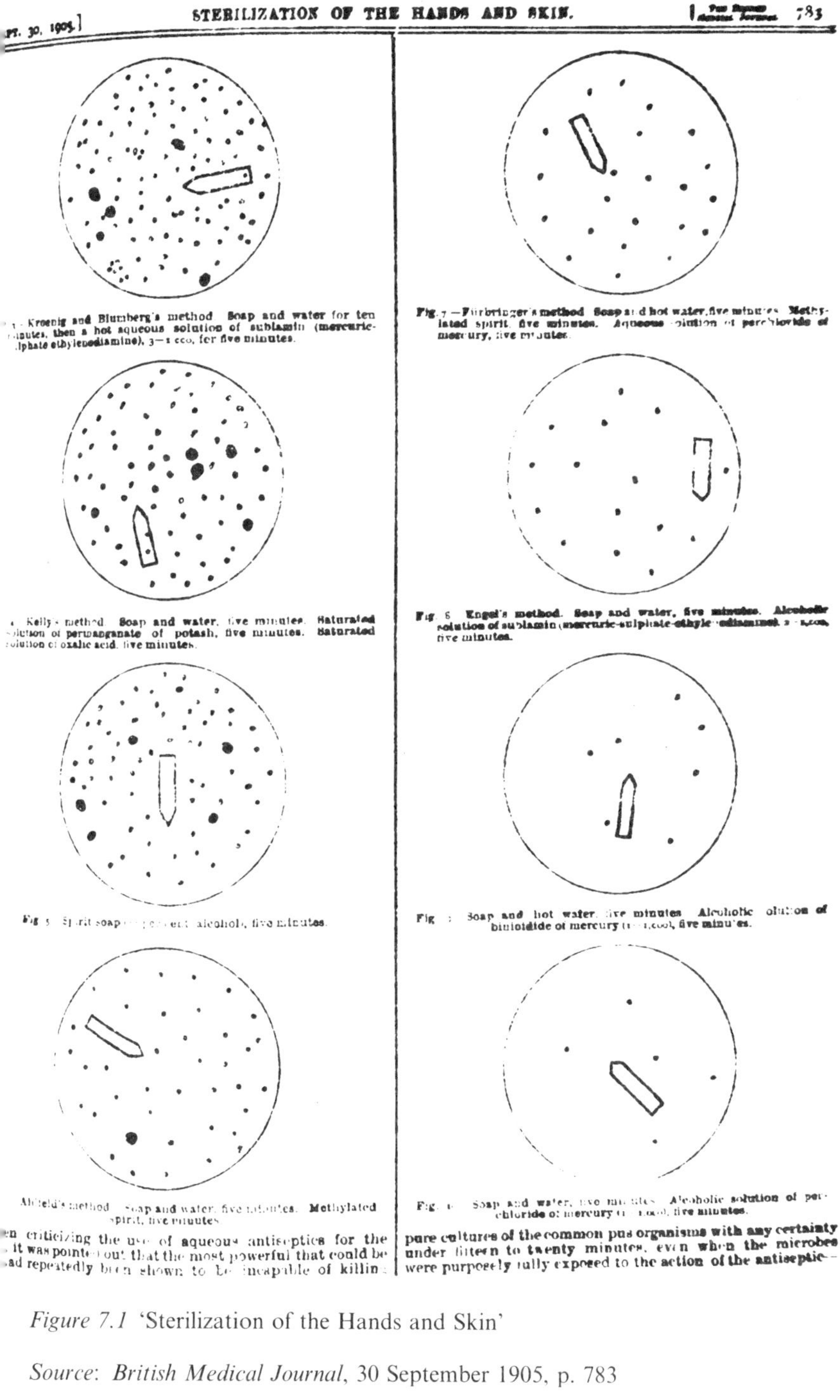

Figure 7.1 'Sterilization of the Hands and Skin'

Source: *British Medical Journal*, 30 September 1905, p. 783

methylated spirit; the hands are then washed in normal saline solution.[51]

The use of various combinations of chemicals and washing was argued over endlessly.

Surgical gloves were tentatively considered from the late 1890s. Cotton, leather and indiarubber gloves were variously tried, with most opinion favouring the latter. Unlike the surgeon's hands, gloves could be sterilised with heat, but most surgeons still remained unconvinced. Even 15 years after they first appeared, gloves were far from universally used. It was argued that they hindered sensitivity, they broke, they were expensive, that surgeons would rely on the sterility of the glove and neglect the process of scrubbing. Gloves were commonly seen as an American and German innovation, which British surgeons were reluctant to endorse fully: 'In England, although our despair in attaining absolute asepsis of the skin may lead us to say, with Lady Macbeth, What, will these hands ne'er be clean? We shall probably continue to prefer the hand to the glove'.[52]

There was always ambiguity about the meaning of gloves. Did they protect the 'clean' patient from the contaminated surgeon, or the 'clean' surgeon from the contaminated patient? Clearly, they could imply both. They created a barrier at the point of contact between two bodies, stopping germs both leaving and entering each body. In turn-of-the-century usage, however, germs were rarely conceptualised as 'flowing' both ways. It is indicative of the surgeon's new vision of himself as contaminated that gloves, when they were used, were nearly always understood to prevent the flow of germs from the dirty practitioner to the clean patient (often articulated as the clean 'wound'). The 'septic case' which reversed this was the exception, abnormal.

In all of this discussion of asepsis, hands, scrubbing and gloves were the major issue, suggesting the extent to which surgical contagion or infection had shifted from being about the air to being about touch and skin. But other parts of the surgeon also came to be scrutinised. A 1911 surgical text had sections on 'care of the operator's mouth, nose and hair'.[53] The surgeon's breath was seen by some to be a possible cause of infection.[54] One mask was designed and advocated so that 'all possible contamination of the wound by such means as dandruff falling from the hair, or perspiration from the brow is avoided'.[55] In other cases, beards were suspect. 'The beard is without doubt guilty of harbouring septic germs', wrote one doctor, who had contrived a 'beardguard'. Another recommended 'that the operator should not wear a mask, but keep his beard moist in sublimate solution during an operation'.[56] Elsewhere, it was the surgeon's more general health which was at issue: 'It is important... for the surgeon to see that he has no carious teeth that have not been stopped, and he should not operate if he be suffering from an influenza cold.'[57] Like the 'new nurse' and the accoucheur earlier in the century, the surgeon himself came to be pathologised.

As I suggested in Chapter 4, the connections between debates on puerperal fever in the middle of the century and early twentieth-century surgical practice are rarely drawn by historians. But the links are striking enough for the genealogy of modern surgery to be rethought. Instead of following an historical and explanatory line from pre-anaesthetic and pre-antiseptic surgery through the antiseptic 'revolution' to asepsis (of course, such a story has already been questioned, but in order to support a different argument), I would suggest an alternative or at least a parallel history which connects obstetric practice to aseptic practice. This is already hinted at with the intermittent inclusion of Semmelweis's work in the genealogy of asepsis and understandings of contact infection.[58] But the connections are far more fundamental. A new story can be told which emphasises the significance of the 'pathologised practitioner' and which connects the crucial marking of the doctor as asymptomatic carrier at both moments.

In the mid nineteenth century, puerperal fever and surgical fever or pyaemia were linked fairly commonly; indeed the suggestion was that puerperal fever was a septic wound infection.[59] This connection was made by Alexander Gordon as early as the late eighteenth century. The early use of antiseptics, even the specific use of carbolic acid, would seem to connect obstetrics with early antiseptic surgery. Yet in surgical discussions at the turn of the century there is barely a mention of what would seem to be important precedents. Surgeons were more intent on writing their story within that of Lister and the antiseptic 'revolution' or within the scientific story of the development of bacteriology since Pasteur.[60] The links between clinical methods of preventing the contagion of puerperal fever and aseptic surgery are readily apparent: concern about hands, instruments, touch and, most importantly, the methods of sanitising or sterilising the body of the surgeon-practitioner. Obstetric practice and aseptic surgical practice problematised direct touch between the patient and practitioner: both, in this sense, were about 'contagion'. While gloves never became a common part of midwifery practice until the twentieth century it was in this context, not in a surgical context, that they were first considered.[61] That both fields of practice – obstetrics and aseptic surgery – debated the use of gloves, the creation of a physical barrier between patient and practitioner, suggests a similar way of thinking about the interaction between the two bodies in any clinical encounter.

Like the accoucheurs in the 1860s, surgical practice around the turn of the century was defined by an intense corporeal self-awareness on the part of the surgeon. The nature of the experiments had shifted from the sort of piecing together of empirical information which characterised the mid nineteenth-century 'case-studies', to laboratory-based bacteriological experiments in which, for example, pieces of the surgeon's skin or nails were removed and placed on agar plates, to see how many colonies of microbes grew. Notwithstanding these sorts of differences, the surgeon's body was unquestionably again the object of scientific study. The surgeon's personal cleanliness was also under question in a way which had earlier characterised the measures recommended for the prevention of puerperal fever. An 1898 surgical text, for example, included a chapter on 'personal asepsis' which reads quite similarly to both mid nineteenth-century instructions to accoucheurs and to texts which instructed nurses in personal hygiene: 'In ordinary polite cleanliness the surgeon should be the model of civilisation. Of course the daily bath must be his habit and his linen unimpeachable always... Hands, of course, are of the greatest importance; nails are best cut short with a regular nail-clipper.'[62] Obstetricians had long been instructing each other to cut their nails, as 'morbid matter' could well be concealed underneath.[63]

Aseptic techniques made sense only through an acceptance of the idea of the doctor as potential contaminator. Yet the moral meanings attached to that potential, that inscription as 'polluter', had changed somewhat from the mid nineteenth century. Turn of the century 'dirtiness' was slightly different, in ways which permitted an easier, less confronting contemplation of oneself as a source of infection. In that germs were everywhere, to have them on and in you was normal: 'The surface of the body is constantly covered with germs and dust, and is also more or less soiled with the various excretions of the body.'[64] It was emphasised in many accounts that even after a rigorous scrubbing, germs were likely to reappear on the hands; the surgeon was truly self-polluting in this sense.[65] This was different to the unidentifiable 'contagion' of the 1860s which attached itself mysteriously to one practitioner but not another. There was a shift towards understanding the normal state of any human body not as a clean, but as contaminated: the normal human body was swarming with germs. It is no coincidence that just when medical practitioners were finally fully implicated as contaminators, moral meanings of that process of contamination began to be minimised, and the process was rewritten as involving the action of morally-neutral or normal germs. There was also an identifiable shift towards constructing germs themselves as active agents of infection. In many texts, germs are discursively invested with an extraordinary independence of action, and the human agent is written out almost completely.[66] There was never entirely a male equivalent to the morally and physically polluted Sairey Gamp.

If the need to create a personal purity was new in male medical culture, spotless cleanliness was already firmly inscribed onto female nursing practice and nursing imagination. Nurses had long been constituted as potential contaminators. If surgeons found themselves with the new imperative to create a 'zone of purity', to use Fox's phrase, a zone which included the embodied surgeon, this was simply not new for women, either in the hospital context or in the domain of public health. For half a century prior to medical acceptance of dirtiness, nurses had been cleaning, sanitising and purifying themselves, their wards, their patients, as well as their cultural representation. The shift from nurses' sanitarian 'purity' to surgeon's aseptic 'sterility' was a complex one, but by making the link, the supposed paradigm shift stemming from germ theories is qualified. Far from 'cleanl[iness] to the point of exquisiteness', to use Nightingale's language, being outdated by 1910, it had assumed a new imperative in a way which incorporated and implicated doctors.[67] Asepsis forced surgeons to adopt a sterility/purity, with its permanent implication of pollution, which had long been central to nursing identity. And so, it was announced in *The Lancet* at the beginning of the new century, 'the surgeon has learnt to be clean'.[68]

Notes and References

Introduction

1. J. Simon, 'Sixth Report of the Medical Officer of the Privy Council' (1863) reprinted in his *Public Health Reports*, vol. 2, J. & A. Churchill, London, 1887, pp. 148–9.
2. The focus of the book is British institutions, debates, events, discourses. Occasionally, I refer to an American or an Australian text, which I see as working well within the same discourse of western medicine.
3. C. E. Rosenberg, 'Florence Nightingale on Contagion: The Hospital as Moral Universe' in Rosenberg (ed.), *Healing and History*, Science History Publications, New York, 1979, p. 124.
4. M. Gatens, *Imaginary Bodies: Ethics, Power and Corporeality*, Routledge, London and New York, 1996, p. 8 ; E. Grosz, *Volatile Bodies: Toward a Corporeal Feminism*, Allen & Unwin, Sydney, 1994, pp. 17–19.
5. 'Nursing Institutions and Hospitals', *The Hospital*, 26 February 1887, p. 375.
6. A recent example is A. Witz, *Professions and Patriarchy*, Routledge, London and New York, 1992.
7. The most frequently cited of these histories include B. Ehrenreich and D. English, *Complaints and Disorders: The Sexual Politics of Sickness*, Compendium, London, 1974; Ann Oakley, 'Wisewoman and Medicine Man: Changes in the Management of Childbirth', in J. Mitchell and A. Oakley (eds), *The Rights and Wrongs of Women*, Penguin, Harmondsworth, 1976, pp. 17–58; L. Gordon, *Woman's Body, Woman's Right*, Penguin, London, 1977.
8. See L. Jordanova, *Sexual Visions: Images of Gender in Science and Medicine between the Eighteenth and Twentieth Centuries*, Harvester Wheatsheaf, London, 1989; M. Poovey, 'Scences of an Indelicate Character: The Medical Treatment of Victorian Women' in *Uneven Developments: The Ideological Work of Gender in Mid-Victorian England*, University of Chicago Press, Chicago, 1988; M. Poovey, *Making a Social Body: British Cultural Formation, 1830–1864*, University of Chicago Press, Chicago and London, 1995; S. Shuttleworth, 'Female Circulation: Medical Discourse and Popular Advertising in the Mid-Victorian Era' in M. Jacobus, E. Fox Keller, S. Shuttleworth (eds), *Body/Politics: Women and the Discourses of Science*, Routledge, London and New York, 1990.
9. L. M. Newman, 'Critical Theory and the History of Women: What's at Stake in Deconstructing Women's History', *Journal of Women's History*, 2, 1991, p. 62; See also J. W. Scott, 'Experience' in J. Butler and J. W. Scott (eds), *Feminists Theorize the Political*, Routledge, London and New York, 1992, pp. 22–40.
10. See L. Jordanova, *Sexual Visions*, pp. 19–42, pp. 87–110; L. Schiebinger, *Nature's Body: Gender in the Making of Modern Science*, Beacon Press, Boston, 1993; E. Grosz, *Volatile Bodies*, pp. 3–24; E. Fox Keller,

150   *Notes and References*

Reflections on Gender and Science, Yale University Press, New Haven and London, 1985, pp. 75–94.
11. M. Foucault, *Discipline and Punish: The Birth of the Prison*, Penguin, Harmondsworth, 1991.

1   Sanitising Spaces: The Body and the Domestic in Public Health

1. B. W. Richardson, 'Woman as a Sanitary Reformer', *Transactions of the Sanitary Institute*, 2, 1880, p. 188.
2. See for example E. H. Ackerknecht, 'Anticontagionism between 1821 and 1867', *Bulletin of the History of Medicine*, 22, 1948, pp. 562–93; R. Cooter, 'Anticontagionism and History's Medical Record' in P. Wright and A. Treacher (eds), *The Problem of Medical Knowledge: Examining the Social Construction of Medicine*, Edinburgh University Press, Edinburgh, 1982; J. M. Eyler, *Victorian Social Medicine: The Ideas and Methods of William Farr*, Johns Hopkins University Press, Baltimore and London, 1979; L. Stevenson, '"Science Down the Drain": On the Hostility of Certain Sanitarians to Animal Experimentation, Bacteriology and Immunology', *Bulletin of the History of Medicine*, 29, 1955, pp. 1–26.
3. P. Williams, 'The Laws of Health: Women, Medicine and Sanitary Reform, 1850–1890' in M. Benjamin (ed.), *Science and Sensibility: Gender and Scientific Enquiry 1780–1945*, Basil Blackwell, Oxford, 1991, pp. 79–80.
4. N. Tomes, 'The Private Side of Public Health: Sanitary Science, Domestic Hygiene, and the Germ Theory, 1870–1900', *Bulletin of the History of Medicine*, 64, 1990, pp. 513–14.
5. For the impact of cholera on the development of the public health movement, see M. Pelling, *Cholera, Fever and English Medicine, 1825–1865*, Oxford University Press, Oxford, 1978.
6. M. Dean, *The Constitution of Poverty: Toward a genealogy of liberal governance*, Routledge, London and New York, 1991, pp. 207–8; C. Lawrence, 'Sanitary Reformers and the Medical Profession in Victorian England' in T. Ogawa (ed.), *Public Health: Proceedings of the Fifth International Symposium on the Comparative History of Medicine East and West*, Saikon Publishing, Tokyo, 1981, pp. 145–46.
7. B. S. Turner, 'The Discourse of Diet' in M. Featherstone, M. Hepworth and B. Turner (eds), *The Body: Social Process and Cultural Theory*, Sage, London, 1991, p. 165.
8. M. Poovey, *Making a Social Body: British Cultural Formation, 1830–1864*, University of Chicago Press, Chicago and London 1995, p. 130.
9. E. Chadwick, 'Address on Public Health', *Transactions of the National Association for the Promotion of Social Science*, 1860, p. 606.
10. M. Foucault, 'The Politics of Health in the Eighteenth Century' in P. Rabinow (ed.), *The Foucault Reader*, Penguin, Harmondsworth, 1991, p. 274.
11. M. Foucault, 'The Politics of Health', p. 277, p. 283.
12. E. Chadwick, *Report on the Sanitary Condition of the Labouring Population of Great Britain* (1842), edited with an introduction by M. W. Finn,

Edinburgh University Press, Edinburgh, 1965, p. 199, passim; See also M. W. Flinn, 'Introduction' to *Report on the Sanitary Condition*, pp. 58–59.
13. A. S. Wohl, *Endangered Lives: Public Health in Victorian Britain*, J. M. Dent, London, 1983, pp. 6–7.
14. Earl of Shaftesbury, 'Address on Public Health', *Transactions of the National Association for the Promotion of Social Science*, 1858, p. 86.
15. C. E. Rosenberg, 'Florence Nightingale on Contagion', p.120; See also F. F. Cartwright, 'Antiseptic Surgery', in F. N. L. Poynter (ed.), *Medicine and Science in the 1860s*, Wellcome Institute for the History of Medicine, London, 1968, p. 81; S. Holton, 'Feminine Authority and Social Order: Florence Nightingale's Conception of Nursing and Health Care', *Social Analysis*, 15, 1984, pp. 60–61.
16. F. Nightingale, *Notes on Nursing: What it is and What it is Not* (1859) Dover Publications, New York, 1969, p. 12.
17. F. F. Cartwright, 'Antiseptic Surgery', p. 87.
18. F. F. Cartwright, 'Antiseptic Surgery', p. 87; F. Galton, *Construction of Hospitals*, Macmillan, London, 1869, p. 13.
19. J. H. Pickford, *Hygiene or Health*, John Churchill, London, 1858, p. 214.
20. Dr Rigby quoted in General Board of Health, *Papers Relating to the Sanitary State of the People of England*, George Eyre and William Spottiswoode, London, 1858, p. xxxviii [original emphasis].
21. F. F. Cartwright, 'Antiseptic Surgery', p. 88.
22. Quoted in C. Lawrence, 'Sanitary Reformers and the Medical Profession', p. 149.
23. Florence Nightingale to Dr John Sutherland, 8 July 1872 in M. Vicinus and B. Nergaard (eds), *Ever Yours, Florence Nightingale: Selected Letters*, Virago, London, 1989, pp. 326–7 [original emphasis and punctuation].
24. E. Fee and D. Porter, 'Public health, preventive medicine and professionalization: England and America in the nineteenth century' in A. Wear (ed.), *Medicine in Society: Historical Essays*, Cambridge University Press, Cambridge, 1992, pp. 249–75.
25. For an analysis of the concept of specific causation see K. Codell Carter, 'The Development of Pasteur's Concept of Disease Causation and the Emergence of Specific Causes in Nineteenth-Century Medicine', *Bulletin of the History of Medicine*, 65, 1991, pp. 528–48.
26. C. E. Rosenberg, 'Florence Nightingale on Contagion', p. 117, p. 124.
27. B. W. Richardson, *Hygeia: A City of Health*, London, 1875, p. 39.
28. See cartoon in *Punch*, 16 December 1875 in which doctors wander through the streets of Hygeia begging and crying 'We Have no Work to Do', reprinted in L. Stevenson, ' "Science Down the Drain" ', p. viii.
29. B. W. Richardson, *Hygeia*, pp. 32–6.
30. J. H. Pickford, *Hygiene or Health*.
31. For discussions of domestic ideology and religion in the English context, see L. Davidoff and C. Hall, *Family Fortunes: Men and Women of the English Middle Class, 1780–1850*, Hutchinson, London and Melbourne, 1987, Part One; C. Hall, 'The Early Formation of Victorian Domestic Ideology' in S. Burman (ed.), *Fit Work for Women*, London and Canberra, 1979; J. Rendall, *The Origins of Modern Feminism*, Macmillan, Basingstoke and London, 1985, pp. 189–230.

32. M. Poovey, *Uneven Developments*, p. 10.
33. E. Chadwick, *Report on the Sanitary Condition*, p. 195.
34. Report of the Physicians and Surgeons at Birmingham, quoted in E. Chadwick, *Report on the Sanitary Condition*, p. 205.
35. Lord Palmerston, Home Secretary, quoted in Wohl, *Endangered Lives*, p. 122.
36. E. Chadwick, *Report on the Sanitary Condition*, p. 174.
37. See for example Galton, *Construction of Hospitals*; D. Galton, *Healthy Dwellings*, Oxford, 1880; F. J. Mouat and H. Saxon Snell, *Hospital Construction and Management*, J.&A. Churchill, London, 1883; F. Oppert, *Hospitals, Infirmaries and Dispensaries*, John Churchill, London, 1867.
38. 'The nurse spoken of is not usually the hired professional person, but the wife, mother, or sister', 'Miss Nightingale's "Notes on Nursing"', *Quarterly Review*, 107, 1860, p. 404.
39. B. W. Richardson, 'Woman as a Sanitary Reformer', p. 188.
40. B. W. Richardson, 'Woman as a Sanitary Reformer', p. 196.
41. H. W. G. Macleod, *Hygiene for Nurses*, Smith Elder, London, 1911, p. 2.
42. 'Second Annual Report of the Ladies' Sanitary Association', *English Woman's Journal*, 3, 1859, pp. 380–7; B. R. Parkes, 'The Ladies' Sanitary Association', *English Woman's Journal*, 3, 1859, pp. 73–85.
43. M. A. Baines, 'The Ladies National Association for the Diffusion of Sanitary Knowledge', *Transactions of the National Association for the Promotion of Social Science*, 1858, p. 531.
44. M. A. Baines, 'The Ladies National Association', p. 531.
45. Anna Jameson, quoted in M. Vicinus, *Independent Women: Work and Community for Single Women, 1850–1920*, Virago, London, 1985, p. 90 [original emphasis].
46. S. R. Powers, 'The Diffusion of Sanitary Knowledge', *Transactions of the National Association for the Promotion of Social Science*, 1860, pp. 715–16; See also P. Williams, 'The Laws of Health', pp. 65–6.
47. W. Cowper, 'Address on Public Health', *Transactions of the National Association for the Promotion of Social Science*, 1859, p. 117.
48. S. R. Powers, 'The Diffusion of Sanitary Knowledge', p. 714.
49. M. A. Baines, 'The Ladies' National Association', pp. 531–2.
50. S. R. Powers, 'The Diffusion of Sanitary Knowledge', p. 715.
51. J. Lewis, *Women and Social Action in Victorian and Edwardian England*, Edward Elgar, Aldershot, 1991, p. 35.
52. *Ladies Health Society of Manchester and Salford*, Richard Gill, Manchester, 1893, p. 5, quoted in J. Lewis, *Women and Social Action*, p. 35.
53. See F. K. Prochaska, *Women and Philanthropy in Nineteenth-Century England*, Oxford University Press, Oxford, 1980, pp. 174–5.
54. E. Yeo, 'Social Motherhood and the Sexual Communion of Labour in British Social Science 1850–1950', *Women's History Review*, 1, 1992, pp. 69–70.
55. For the race politics at work in Nightingale's world-view, see her *Note on the Aboriginal Races in Australia*, London, 1865; F. Nightingale, 'How Some People Have Lived, and not died in India', *Transactions of the National Association for the Promotion of Social Science*, 1873, pp. 463–74.

Notes and References 153

56. M. Poovey, *Uneven Developments*, p. 166; See also, S. Holton, 'Feminine Authority and Social Order', pp. 59–72;
57. S. Powers, 'The Details of Woman's Work in Sanitary Reform', *English Woman's Journal*, 3, 1859, p. 217.
58. S. Powers, *Remarks on Women's Work in Sanitary Reform*, Ladies Sanitary Association, n.d., pp. 3–4 [original emphasis]. For a similar argument that sanitary reform necessarily involved the work of both women and men, see B. R. Parkes, 'The Ladies' Sanitary Association', pp. 73–85.
59. J. Simon, 'Sixth Report of the Medical Officer of the Privy Council', p. 149.
60. A 'lady informant' quoted in E. Chadwick, *Report on the Sanitary Condition*, p. 195.
61. E. Chadwick, *Report on the Sanitary Condition*, p. 196.
62. F. Oppert, *Hospitals, Infirmaries and Dispensaries*, p. 27.
63. F. Nightingale, 'Health Teaching in Towns and Villages', in L. Seymer (ed.), *Selected Writings of Florence Nightingale*, Macmillan, London, 1954, p. 388; See also N. Tomes, 'The Private Side of Public Health', pp. 522–5.
64. D. Armstrong, 'Public Health Spaces and the Fabrication of Identity', *Sociology*, 27, 3, 1993, p. 396.
65. B. W. Richardson, 'Woman as a Sanitary Reformer', pp. 198–99.
66. 'Hospital Hygiene', *The Lancet*, II, 1864, p. 498.
67. *The House We Live In*, Maryborough, 1874, p. 1.
68. F. Nightingale, 'Health Teaching in Towns and Villages', p. 388.
69. F. Nightingale, *Notes on Nursing*, p. 94.
70. A. Thomson, *Lecture on Sanitary Reform*, George and Robert King, Aberdeen, 1860, p. 15.
71. B. W. Richardson, 'Woman as a Sanitary Reformer', pp. 195–96.
72. M. Douglas, *Purity and Danger: An Analysis of the Concepts of Pollution and Taboo* (1966) Routledge, London and New York, 1994, p. 2, p. 36.
73. W. Cowper, 'Address on Public Health', *Transactions of the National Association for the Promotion of Social Science*, 1859, pp. 106–07.
74. R. J. Mann, *Domestic Economy and Household Science*, Edward Stanford, London, 1878, p. 256.

2 Female Bodies at Work: Narratives of the 'Old' Nurse and 'New' Nurse

1. 'The best for the purpose are... taken from the recruits of upper class domestic servants of good health, intelligence and education... At present I have plenty of lady applicants but alas the right sort do not always come forward. Small delicate people are not suitable however willing they are to work. It needs a certain amount of presence and a good practical knowledge of the work as well as real goodness and a high moral tone', Miss Williams to Miss L. M. Hubbard, 15 August 1877, Autograph Letters Collection, Fawcett Library; See also R. Dingwall *et al.*, *An Introduction to the Social History of Nursing*, Routledge, London, 1988, p. 33.

2. J. Jebb, 'Statement of the Appropriation of the Nightingale Fund', *Transactions of the National Association for the Promotion of Social Science*, 1863, p. 643.
3. Other nursing sisterhoods and institutions which were contracted by hospitals to manage and deliver nursing services included the Deaconesses of the London Diocesan for the Great Northern Hospital, the Evangelical Protestant Deaconesses' Institution at Tottenham Green for the Perth and Sunderland Hospitals, the British Nurses' Association for the Royal Free Hospital. See J. C. Steele, 'Nursing and Nursing Institutes', *The Sanitary Record*, January 1875, pp. 4–6. For a full analysis of the Anglican sisterhoods see J. Moore, *A Zeal for Responsibility: The Struggle for Professional Nursing in Victorian England, 1868–1883*, University of Georgia Press, Athens and London, 1988; M. Vicinus, *Independent Women*, pp. 46–84;
4. St John's House and Sisterhood, *Rules and Regulations*, Harrison & Sons, London, 1867, p. 7.
5. A. Summers, *Angels and Citizens: British Women as Military Nurses, 1854–1914*, Routledge and Kegan Paul, London & New York, 1988, p. 20; J. Moore, *A Zeal for Responsibility*, p. 3.
6. M. Vicinus, *Independent Women*, p. 89.
7. M. Baly, *Florence Nightingale and the Nursing Legacy*, Croom Helm, London, 1986, pp. 37–8.
8. 'Regulations as to Training of Special Probationers', n.d. (?1867) Nightingale Training School Records, GLRO, HI/ST/NTS/A2/3.
9. 'Regulations as to the Training of Hospital Nurses under the Nightingale Fund', 1872, Nightingale Training School Records, GLRO, H1/ST/NTS/A2/2.
10. 'Method of Training Nurses at St Thomas's and King's College Hospitals' from 'Report on Cubic Space in Metropolitan Workhouses, Blue Book, 1867', Nightingale Fund Collection, GLRO, A/NFC/95/10.
11. Countess de Viesca, 'On Sanitary House Management, *Transactions of the Sanitary Institute*, 7, 1885–6, p. 129.
12. For succinct summaries of the range of new initiatives in hospital, district, workhouse and religious nursing, see B. Abel-Smith, *A History of the Nursing Profession*, Heinemann, London, 1960, pp. 17–35; H. C. Burdett, *Hospitals and Asylums of the World*, vol. 3, Churchill, London, 1893, pp. 248–53; A. Summers, *Angels and Citizens*, pp. 13–28.
13. See for example B. Abel-Smith, *A History of the Nursing Profession*, p. 17; B. Abel-Smith, *The Hospitals 1800–1948*, Heinemann, London, 1964, pp. 43–5; C. Helmstadter, 'Robert Bentley Todd, Saint John's House, and the Origins of the Modern Trained Nurse', *Bulletin of the History of Medicine*, 67, 1993. p. 285.
14. See for example F. Nightingale, 'Notes on the Sanitary Conditions of Hospitals', *Transactions of the National Association for the Promotion of Social Science (TNAPSS)* 1858, pp. 62–82; Mrs J.N. Higgins, 'On the Improvement of Nurses in Country Districts', *TNAPSS*, 1861, pp. 572–76; F. Nightingale, 'Hospital Statistics and Hospital Plans', *TNAPSS*, 1861, pp. 554–60; J. S. Howson, 'The Official Employment of Women in Works of Charity', *TNAPSS*, 1862, pp. 780–83; E. Garrett, 'Hospital

Nursing', *TNAPSS*, 1866, 472–8 and response pp. 588–94; M. Merryweather, 'The Training of Educated Women for Superintendents', *TNAPSS*, 1867, p. 452.
15. M. Dean and G. Bolton, 'The Administration of Poverty and the Development of Nursing Practice in Nineteenth-Century England' in C. Davies (ed.), *Rewriting Nursing History*, Croom Helm, London, 1980.
16. M. Foucault, 'The Politics of Health in the Eighteenth Century', p. 280, p. 283.
17. Lady Strangford, *Hospital Training for Ladies: An Appeal to the Hospital Boards in England*, Harrison & Sons, London, 1874, p. 13.
18. A. Jameson, *The Communion of Labour*, London, 1855, quoted in Prochaska, *Women and Philanthropy*, p. 174.
19. See M. Lonsdale, 'The Present Crisis at Guy's Hospital', *Nineteenth Century*, 7, 1880, pp. 677–84; 'Doctors and Nurses', *Nineteenth Century*, 7, 1880, pp. 1089–108. The ongoing difficulties of the St John's House Sisters are admirably dealt with in J. Moore, *A Zeal for Responsibility*.
20. 'Second Annual Report of the Ladies' Association for the Diffusion of Sanitary Knowledge', *English Woman's Journal*, 3, 1859, p. 382.
21. Mary Anne Baines to Florence Nightingale, 19 December 1859, Nightingale Papers, BL Add. MSS. 45, 793 ff. 196–8 [original emphasis].
22. Queen Victoria Jubilee Institute for Nurses, Medical and Sanitary Sub-Committee Minutes, 8 July 1890, held by the Queen's Nursing Institute, London.
23. See for example M. Loane and H. Bowers, *The District Nurse as Health Missioner*, Women's Printing, London, n.d. [?1908].
24. A. Summers, 'The Mysterious Demise of Sarah Gamp: The Domiciliary Nurse and her Detractors ', *Victorian Studies*, 32, 1989, p. 365.
25. M. Trench, 'Sick-Nurses', *Macmillan's Magazine*, 34, 1876, p. 425; See also Lonsdale, 'The Present Crisis at Guy's Hospital', p. 678; 'Hospital Nurses as they are and as they ought to be', *Frasers Magazine*, 37, 1848, p. 540.
26. Miss Hamilton, *A Course of Four Lectures on Sick-Nursing*, Clunes, Guardian and Gazette, 1886, p. 4.
27. F. Milford, *An Australian Handbook of Obstetric Nursing*, Angus & Robertson, Sydney, 1896, p. viii.
28. *Nurses' Chronicle*, 1, September 1887, p. 2.
29. F. Nightingale, 'Nursing the Sick' in L. Seymer (ed.), *Selected Writings of Florence Nightingale*, Macmillan, London, 1954, p. 344.
30. B. W. Richardson, *Hygeia*, p. 35.
31. F. Churchill, *A Manual for Midwives and Monthly Nurses*, Longman, London, 1856, p. 2.
32. B. R. Parkes, 'At a Nurses' Training School', *Alexandra Magazine*, February 1865, p. 70.
33. Lucy Osburn to Florence Nightingale, 4 December 1868, BL Add. MSS 47,757, ff. 96–7.
34. Mary Barker to Florence Nightingale, 30 May 1868, BL Add. MSS 47, 757, f. 235.
35. Haldane Turriff to Florence Nightingale, 29 January 1869, BL Add. MSS. 47, 757, ff. 248–49.

36. F. G. Holden, *Her Father's Darling and Other Child Pictures*, Turner and Henderson, Sydney, 1887, pp. 76–7.
37. F. G. Holden, 'Petticoat Government', *Sydney Quarterly Magazine*, 1, 1884, p. 271.
38. Cited in E. G. Fenwick, 'Nurses à la Mode: A Reply to Lady Priestley', *Nineteenth Century*, 41, 1897, p. 326.
39. *Nursing Record and Hospital World*, 12, 1894, p. 5.
40. *The Afternoon of Unmarried Life* (1858) quoted in M. Vicinus, *Independent Women*, p. 13.
41. E. Garrett, 'Hospital Nursing' (1866) reprinted in C. A. Lacey (ed.) *Barbara Leigh Smith Bodichon and the Langham Place Group*, Routledge & Kegan Paul, New York and London, 1987, p. 445.
42. E. Lückes, *Lectures on General Nursing*, Kegan Paul, Trench & Co., London, 1884, pp. 225–6 [original emphasis].
43. M. Douglas, *Purity and Danger*, p. 116; See also D. Lupton, *Medicine as Culture: Illness, Disease and the Body in Western Societies*, Sage, London, 1994, pp. 20–49.
44. See for example L. Jordanova, *Sexual Visions*, passim; M. Poovey, *Uneven Developments*, pp. 24–50; S. Shuttleworth, 'Female Circulation: Medical Discourse and Popular Advertising in the Mid-Victorian Era', pp. 47–68.
45. L. Jordanova, *Sexual Visions*, p. 135.
46. See for example S. Bell, *Reading, Writing and Rewriting the Prostitute Body*, Indiana University Press, Bloomington and Indianapolis, 1994, pp. 40–72; S. K. Kent, *Sex and Suffrage in Britain*, Princeton University Press, Princeton, 1987, pp. 60 ff; L. Nead, *Myths of Sexuality: Representations of Women in Victorian Britain*, Basil Blackwell, Oxford, 1988.
47. See M. Spongberg, 'The Sick Rose: Constructing the Body of the Prostitute in Nineteenth Century British Medical Discourse', PhD thesis, University of Sydney, 1992.
48. M. Poovey, *Making a Social Body*, pp. 92–3; See also A. Corbin, 'Commercial Sexuality in Nineteenth-Century France: A System of Images and Regulation' in C. Gallagher and T. Laqueur (eds), *The Making of the Modern Body*, University of California Press, Berkeley, 1987.
49. L. Davidoff and C. Hall, *Family Fortunes*, p. 90.
50. J. Simon, 'Sixth Report of the Medical Officer of the Privy Council', p. 149.
51. D. Lupton, *Medicine as Culture*, p. 141.
52. For an important recent analysis see A. Anderson, *Tainted Souls and Painted Faces: The Rhetoric of Fallenness in Victorian Culture*, Cornell University Press, Ithaca and London, 1993.
53. In 1864 hospital gangrene was described as 'the offspring of an unclean embrace that sullies the virgin purity of the blood by a detestable impregnation; a mysterious, propagable, depraved, terrible something, we know not what', quoted in C. E. Rosenberg, 'Florence Nightingale on Contagion', p. 122.
54. *The Australian Concise Oxford Dictionary*, Oxford University Press, Melbourne, 1987.

Notes and References 157

55. J.A. Hornsby and R.E. Schmidt, *The Modern Hospital: Its Inspiration, Its Architecture, Its Equipment, Its Operation*, W. B. Saunders, London and Philadelphia, 1913, p. 325.
56. E. Grosz, *Volatile Bodies*, p. 204.

3 'Disciplines of the Flesh': Sexuality, Religion and the Modern Nurse

1. M. Foucault, *Discipline and Punish*, Penguin, Harmondsworth, 1991.
2. B. S. Turner, *Medical Power and Social Knowledge*, Sage, London and Beverly Hills, 1987, p. 20; See also M. Foucault, *The History of Sexuality, An Introduction*, Penguin, Harmondsworth, 1987, p. 67.
3. B. S. Turner, *Regulating Bodies*, p. 22; B. S. Turner, 'Recent Theoretical Developments in the Sociology of the Body', *Australian Cultural History*, 13, 1994, p. 27.
4. B. S. Turner, 'The Discourse of Diet' in M. Featherstone, M. Hepworth and B. Turner (eds), *The Body: Social Process and Cultural Theory*, Sage, London, 1991, p. 158.
5. See for example, B. Abel-Smith, *A History of the Nursing Profession*, p. 22; V. Bullough and B. Bullough, 'Nursing, Sexual Harassment, and Florence Nightingale: Implications for Today' in V. Bullough et al. (eds), *Florence Nightingale and Her Era: A Collection of New Scholarship*, Garland, New York and London, 1990, pp. 176–80; C. Webb, *Sexuality, Nursing and Health*, John Wiley & Sons, 1985, pp. 130–1.
6. B. S. Turner, 'Weber on Medicine and Religion' in his *For Weber, Essays on the Sociology of Fate*, Routledge & Kegan Paul, Boston, London and Henley, 1981, pp. 177–99.
7. M. Weber, *Economy and Society*, quoted in B. S. Turner, *For Weber*, p. 179.
8. B. S. Turner, *For Weber*, p. 181.
9. B. S. Turner, *The Body and Society: Explorations in Social Theory*, Basil Blackwell, Oxford, 1984, p. 216.
10. M. Foucault, *Discipline and Punish*, p. 138.
11. M. Foucault, *Discipline and Punish*, p. 178.
12. M. Foucault, *Discipline and Punish*, p. 140, p. 143.
13. L. L. Dock, 'The Relation of Training Schools to Hospitals' in I. A. Hampton (ed.), *Nursing of the Sick – 1893*, McGraw-Hill, 1949, p. 16.
14. E. C. Laurence, *Modern Nursing in Hospital and Home*, Scientific Press, London, 1907, p. 4; M. Vicinus, *Independent Women*, p. 88, pp. 92–3.
15. M. Poovey, *Uneven Developments*, p. 170.
16. J. Bethke Elshtain, *Women and War*, Harvester Press, Brighton, 1987, p. 4.
17. A. Summers, *Angels and Citizens*, passim.
18. Harriet Martineau, Obituary of Florence Nightingale, quoted in M. Poovey, 'A Housewifely Woman: The Social Construction of Florence Nightingale' in her *Uneven Developments*, p. 165.
19. S. Holland, *A Talk to the Nurses of the London Hospital*, Whitehead, Morris & Co, London, n.d., pp. 17–18 [original emphasis].

20. 'Hospital Discipline and Ethics', *Australasian Nurses' Journal*, 6, 1908, pp. 18–19.
21. M. Foucault, *Discipline and Punish*, p. 139.
22. E. C. Laurence, *Modern Nursing in Hospital and Home*, pp. 7–8.
23. M. Foucault, *Discipline and Punish*, p. 166.
24. H. C. Barclay, 'Discipline and Etiquette', *Australasian Nurses' Journal*, 10, 1912, p. 231.
25. E. Lückes, *Lectures on General Nursing*, pp. 16–19 [original emphasis].
26. M. Foucault, *Discipline and Punish*, p. 149.
27. The 'Day and Night Table', Appendix 4 in *Method of Training Nurses at St Thomas's and King's College Hospitals (Under the Nightingale Fund)*, Nightingale Fund Collection, GLRO, A/NFC/95/10; See also S. Holton, 'Feminine Authority and Social Order', p. 63.
28. See for example, E. Lückes, *Lectures on General Nursing*, pp. 12–13.
29. M. Foucault, *Discipline and Punish*, pp. 170–1.
30. A. Roberts, 'On the Hospital Requirements of Sydney', *Transactions of the Royal Society of New South Wales*, 2, 1868, pp. 28–9.
31. F. Nightingale, 'Subsidiary Notes as to the Introduction of Female Nursing into Military Hospitals', quoted in M. Poovey, *Uneven Developments*, p. 181.
32. A. Nutting, 'The Social Work of Nurses', *Australasian Nurses' Journal*, 10, 1912, reprinted from the *British Journal of Nursing*.
33. For one analysis of the influence of religious debate on nursing see P. Williams, 'Religion, respectability and the origins of the modern nurse' in R. French and A. Wear (eds), *British Medicine in an Age of Reform*, Routledge, London and New York, 1991, pp. 239–41.
34. P. Williams, 'Religion, respectability', p. 244.
35. A. Summers, *Angels and Citizens*, p. 43.
36. M. Stanley, *Hospitals and Sisterhoods*, John Murray, London, 1855, p. 1.
37. A. Jameson, *Sisters of Charity, Catholic and Protestant, Abroad and at Home*, Longman, Brown, Green & Longmans, London, 1855, pp. 5–7.
38. C. Haddon, 'Nursing as a Profession for Ladies', *St Paul's Magazine*, August 1871, pp. 458–61.
39. See for example L. Dock and M. A. Nutting, *A History of Nursing*, vol. 1, G. P. Putnam & Sons, New York and London, 1907; S. Tooley, *The History of Nursing in the British Empire*, Bousefield, London, 1906.
40. S. Holland, *A Talk to the Nurses of the London Hospital*, p. 4, p. 11.
41. M. Vicinus, *Independent Women*, p. 90.
42. 'One has passed away without noise, without crown or sceptre of martyrdom... Her kingdom was that of the sick. No public press heroine was she; yet countless sick will bless the work of their unknown benefactress.' F. Nightingale, 'A Silent Heroine', *Dawn*, 5, 1893, p. 13, reprinted from the *British Medical Journal*. See also the preface by Nightingale in *Memorials of Agnes Elizabeth Jones, by her Sister*, Strahan & Co., London, 1871; M. Vicinus, *Independent Women*, p. 90.
43. See M. Poovey, *Uneven Developments*, p. 168, n. 9.
44. Mrs Ormiston Chant, 'Christianity and Nursing', *The Nursing Record and Hospital World*, 9 February 1893, p. 79.

45. For example Nightingale described the Anglican sisters of St John's House and the Catholic sisters who accompanied her to the Crimea, respectively as 'the 6 most perfectly useless specimens of the animal creation I have ever seen...the converting...childish fools of nuns.' Florence Nightingale to Henry Bonham Carter, 2 December 1867, BL Add. MSS 47, 715, f. 122.
46. M. Poovey, *Uneven Developments*, p. 198.
47. Lucy Osburn to Florence Nightingale, 16 June 1869, BL Add MSS 47, 757, f. 118.
48. Lucy Osburn to Florence Nightingale, 12 May 1873, BL Add. MSS 47, 757, f. 144.
49. Mary Barker to Florence Nightingale, 30 May 1868, BL Add. MSS 47, 757, ff. 235–6.
50. Haldane Turriff to Florence Nightingale, 21 February 1869, BL Add. MSS 47,757, f. 258; Evidence of Sister Haldane, Report of the Sub-Committee to Inquire into the Allegations of the *Protestant Standard*, New South Wales Legislative Assembly, *Votes and Proceedings*, vol. 4, 1870–1, p. 8.
51. M. Vicinus, *Independent Women*, p. 88, p. 92.
52. For a discussion of Weber's arguments about Protestantism, asceticism and salvation, see B. S. Turner, *Regulating Bodies: Essays in Medical Sociology*, Routledge, London and New York, 1992, pp. 197–200.
53. H. Morten, *Sketches of Hospital Life*, Sampson, Low, Marston, Searle & Rivington, London, 1888, p. 7.
54. C. J. Wood, *A Handbook of Nursing for the Home and the Hospital*, Cassell & Co., London and Melbourne, n.d. (189?), p. 11, p. 39.
55. E. Lückes, *Lectures on General Nursing*, p. 172, p. 227.
56. M. Trench, 'Sick-Nurses', *Macmillans Magazine*, 34, 1876, p. 423.
57. 'Uniform', *The Nursing Record and Hospital World*, 2 February 1895, p. 80.
58. A. Munro, *The Science and Art of Nursing the Sick*, James Maclehouse, Glasgow, 1873, p. 104.
59. J. Blair, *Opening Address to Nurses*, George Robertson, Sydney, 1880, p. 10.
60. 'The Perfect Nurse – Some Rules of Prudence which Deserve Attention', *Australasian Nurses' Journal*, 4, 1906, p. 373.
61. 'A Code of Ethics', *Australasian Nurses' Journal*, 6, 1908, p. 65 reprinted from the *British Journal of Nursing*.
62. B. S. Turner, *The Body and Society*, pp. 63–4.
63. M. Gatens, 'Towards a Feminist Philosophy of the Body' in B. Caine *et al.*, (eds), *Crossing Boundaries: Feminisms and the Critique of Knowledges*, Allen & Unwin, Sydney, 1988, p. 60; See also G. Lloyd, *The Man of Reason*, London, Methuen, 1984; C. Pateman, *The Disorder of Women*, Polity Press, Cambridge, 1989, pp. 17–29.
64. M. Poovey, *Uneven Developments*, p. 11.
65. E. Gordon Fenwick, 'Nurses à la Mode: a Reply to Lady Priestley', *Nineteenth Century*, 41, 1897, p. 332. Another article described a 'burlesque at one of the London theatres, where Nurses are represented taking part in very questionable dances and songs. A band of Nurses under the

name of the Hot Cross Bun Brigade, in uniform, and with the large red cross on the arm, are shown... "Miss Nightingale," who is at the head of the band, is an accomplished skirt-dancer, and figures throughout the whole piece as a person of most frivolous and objectionable type.' 'Nurses in Burlesque', *Nursing Record and Hospital World*, 9 June 1894, p. 370.
66. *Nursing Record and Hospital World*, 8 June 1893, p. 287.
67. 'Outdoor Uniform for Nurses', *Australasian Nurses' Journal*, 4, 1906, p. 316.
68. J. Bell, *Notes on Surgery for Nurses*, Oliver and Boyd, Edinburgh, 1899, p. 185.
69. Probationer, letter to editor *Australasian Nurses' Journal*, 1, 1903, p. 65.
70. 'Nurses Uniforms Worn in the Street', *Una*, 4, 1906, p. 31.
71. M. Poovey, *Uneven Developments*, p. 177.
72. J. A. Hornsby and R. E. Schmidt, *The Modern Hospital*, p. 325.
73. B. S. Turner, *Regulating Bodies*, p. 197.

4 Pathologising the Practitioner: Puerperal Fever in the 1860s

1. The term 'obstetrician' was not in common usage in the 1860s. When employed it usually referred to a specialist who would have had some attachment to a lying-in hospital or ward. The term 'accoucheur' was far more common – a man-midwife or even general practitioner who worked largely in the field of childbirth. Some specialists labelled themselves 'physician-accoucheur'. Most accoucheurs referred to their practice as 'midwifery' not 'obstetrics'. Thus, 'midwifery' textbooks published even in the 1880s and 1890s were directed to doctors, not midwives.
2. See M. Poovey, *Uneven Developments*, pp. 24–50.
3. A. Rich, *Of Woman Born: Motherhood as Experience and Institution*, Virago, London, 1977, p. 153; A. Rubenstein also emphasises medical men's denial of their role in spreading puerperal fever in 'Subtle Poison: The Puerperal Fever Controversy in Victorian Britain', *Historical Studies*, 20, 1983, pp. 420–38.
4. M. Poovey, *Uneven Developments*, p. 41.
5. O. Moscucci, *The Science of Woman: Gynaecology and Gender in England, 1800–1929*, Cambridge University Press, Cambridge, 1990, p. 60.
6. O. Moscucci, *The Science of Women*, pp. 64–7.
7. O. Moscucci, *The Science of Woman*, p. 71.
8. C. Blake, *The Charge of the Parasols: Women's Entry to the Medical Profession*, Women's Press, London, 1990, p. 180.
9. See J. M. Benn, *Predicaments of Love*, Pluto Press, London, 1992, pp. 116–17; J. Donnison, *Midwives and Medical Men*, Heinemann, London, 1977, pp. 74–5.
10. Female Medical Society, *Report of the Third Annual General Meeting*, 1867, p. 6.
11. 'Female Medical Society', 1864, pamphlet in Women's Medical Education Pamphlets, Fawcett Library.

Notes and References

12. J. M. Benn, *Predicaments of Love*, p. 116; C. Blake, *The Charge of the Parasols*, pp. 79–83.
13. Female Medical Society, *Report of the Third Annual General Meeting*, 1867, p. 11.
14. 'Female Medical Society Prospectus' (?1872) enclosed in Nightingale Correspondence, GLRO H1/ST/NC3/C/72/2.
15. Female Medical Society, *Report of the Third Annual General Meeting*, 1867, p. 4.
16. 'Charge Against a Midwife', *The Lancet*, I, 1863, p. 73.
17. See for example H. Bennett, 'Women as Practitioners of Midwifery', *The Lancet*, I, 1870, p. 887.
18. J. S. Mill to Dr Edmunds, 28 August 1865, printed in James Edmunds, *The Introductory Address Delivered for the Female Medical Society*, n.d., bound in Women's Medical Education Pamphlets, Fawcett Library.
19. M. G. Fawcett, 'The Medical and General Education of Women', *Fortnightly Review*, 4, 1868, 554–55.
20. F. B. Smith, *Florence Nightingale: Reputation and Power*, Croom Helm, London and Canberra, 1982, p. 162.
21. 'Second Annual Meeting of the Female Medical Society', *Victoria Magazine*, August 1866, p. 342; 'Mortality in Childbirth', *The Times*, 10 October 1865.
22. J. Edmunds, 'The Introductory Address Delivered for the Second Session of the Female Medical Society', *Victoria Magazine*, November 1865, p. 52.
23. *Medical Times and Gazette*, 25 August 1866, pp. 195–6, pp. 208–9; See also J. M. Benn, *Predicaments of Love*, pp. 118–21.
24. Mrs Thorne, 'Second Essay' in *Prize Essays on 'Search'*, L. Booth, London, 1867, p. 9; See also J. Donnison, *Midwives and Medical Men*, pp. 82–3.
25. 'An Old Woman', *Prize Essays on 'Search'*, L. Booth, London, 1867, p. 22.
26. See W. Talley, *He, or Man-Midwifery and the Results: or, Medical Men in the Criminal Courts*, Job Caudwell, London, 1863.
27. 'Female Medical Society', *Victoria Magazine*, 9, 1867, p. 209.
28. C. Lawrence, 'Incommunicable Knowledge: Science, Technology and the Clinical Art in Britain 1850–1914', *Journal of Contemporary History*, 20, 1985, pp. 503–5.
29. 'Kathar', 'Murder of the Innocents', *Prize Essay on 'Search'*, L. Booth, London, 1867, p. 11.
30. W. Talley, *He, or Man-Midwifery*, p. 1.
31. W. Talley, *He, or Man-Midwifery*, p. 33 [original emphasis].
32. See for example the chapter 'Hands of Flesh, Hands of Iron' in A. Rich, *Of Woman Born*.
33. W. Tyler Smith, 'A Course of Lectures on the Theory and Practice of Obstetrics', *The Lancet*, I, 1856, p. 675.
34. C. Lawrence, 'Incommunicable Knowledge', p. 505, p. 517.
35. C. Lawrence, 'Incommunicable Knowledge', p. 505.
36. W. Tyler Smith, 'A Course of Lectures on the Theory and Practice of Obstetrics', *Lancet*, I, 1856, p. 676.
37. M. Douglas, *Purity and Danger*, p. 122.

38. J. Edmunds, 'Introductory Address Delivered for the Female Medical Society', p. 6; See also J. Edmunds, 'Female Medical Society and Mortality in Childbirth', *Medical Times and Gazette*, 25 August 1866, p. 208.
39. I. Thorne, 'Second Essay', p. 13.
40. E. W. Murphy, *Lectures on the Principles and Practice of Midwifery*, Walton and Maberly, London, 1862, p. 682.
41. *The Lancet*, I, 1869, p. 819.
42. *The Lancet*, II, 1869, p. 344.
43. R. Barnes, 'Puerperal Fever', *The Lancet*, I, 1865, p. 170.
44. Medical Board Minutes, 8 January 1868, King's College Hospital Archives, KH/MB/M3.
45. See Florence Nightingale, *Introductory Notes on Lying-In Institutions*, Longmans, Green & Co., London, 1871, p. 32, p. 73; C. Rowling, 'The History of the Florence Nightingale Lying-In Ward, King's College Hospital', *Transactions of the Obstetrical Society of London*, 10, 1869, p. 55.
46. C. Rowling, 'The History of the Florence Nightingale Lying-In Ward', pp. 52–4; Priestley, 'On the Closing of the Nightingale Ward, King's College Hospital', *Medical Times and Gazette*, 1868, p. 90.
47. M. Pelling, *Cholera, Fever and English Medicine*, pp. 297–99.
48. J. M. Good, *The Study of Medicine*, London, 1825 cited in C. Hamlin, 'Predisposing Causes and Public Health in Early Nineteenth-Century Medical Thought', *Social History of Medicine*, 5, 1992, p. 47. For specific causes, see K. Codell Carter, 'The Development of Pasteur's Concept of Disease Causation', pp. 528–48.
49. K. Codell Carter, 'Ignaz Semmelweis, Carl Mayrhofer, and the Rise of Germ Theory', *Medical History*, 29, 1985, pp. 33–4.
50. See for example, J. Winn, *Outlines of Midwifery*, Longmans, Brown, Green & Longman, London, 1854, pp. 298 ff; Dr Snow Beck, 'On a Case of Puerperal Fever', *The Lancet*, II, 1867, p. 805.
51. W. Tyler Smith, 'Puerperal Fever', *The Lancet*, II, 1856, p. 505.
52. F. Churchill, *A Manual for Midwives and Monthly Nurses*, Longman & Co., London, 1856, p. 2, p. 32.
53. For the importance of washing with water usshered in with a bacteriological world, see G. Vigarello, *Concepts of Cleanliness*, Cambridge University Press, Cambridge, 1992, pp. 202 ff.
54. R. Barnes, 'Puerperal Fever', *The Lancet*, II, 1865, p. 532.
55. R. Barnes, 'Puerperal Fever', *The Lancet*, II, 1865, p. 614.
56. 'Epidemic of Puerperal Fever at Maidenhead', *The Lancet*, II, 1865, p. 602.
57. Dr Snow Beck, 'On a Case of Puerperal Fever', *The Lancet*, II, 1867, p. 805.
58. 'Obstetrical Society of London', *The Lancet*, I, 1870, p. 485. See also *The Lancet*, I, 1869, p. 819.
59. R. Barnes, 'Puerperal Fever', *The Lancet*, I, 1865, p. 141.
60. R. Barnes, *On Antiseptic Midwifery and Septicaemia in Midwifery*, W. M. Wood, New York, 1882, p. 10.
61. R. Barnes, 'Lectures on Puerperal Fever,' *The Lancet*, II, 1865, p. 613.

62. T. Snow Beck, 'On Puerperal Fever', *The Lancet*, I, 1865, p. 340.
63. See E. Grosz, *Volatile Bodies*, p. 203.
64. D. Armstrong, *Political Anatomy of the Body: Medical Knowledge in Britain in the Twentieth Century*, Cambridge University Press, Cambridge, 1983, p. 9; D. Armstrong, 'Public Health Spaces and the Fabrication of Identity', *passim*.
65. 'Victoria Discussion Society', *Victoria Magazine*, 14, 1870, p. 514.
66. Florence Nightingale to Harriet Martineau, 24 September 1861, Nightingale Papers, BL Add. Mss 45, 788 ff. 131–2.
67. Cited in *The Lancet*, II, 1856, p .446.
68. See A. Bashford, 'Separatist Health: Changing Meanings of Women's Hospitals, c. 1870–1920' in L. R. Furst (ed.), *Women Healers and Physicians: Climbing a Long Hill*, University Press of Kentucky, Lexington, 1997.
69. W. Tyler Smith, 'Puerperal Fever', *The Lancet*, II, 1856, pp. 504–5.
70. F. Churchill, *On the Theory and Practice of Midwifery*, Henry Renshaw, London, 1866, pp. 682–90.
71. F. Churchill, *On the Theory and Practice of Midwifery*, pp. 687–8 [original emphasis].
72. F. Churchill, *On the Theory and Practice of Midwifery*, pp. 691–92.
73. Dr Meigs quoted in F. Churchill, *On the Theory and Practice of Midwifery,* p. 687.
74. Wynn Williams, 'On Puerperal Fever', *The Lancet*, I, 1870, p. 231.
75. See C. Hamlin, 'Predisposing Causes and Public Health', pp. 51–2.

5 Feminising Medicine: The Gendered Politics of Health

1. 'Female Medical Society', *Victoria Magazine*, 9, 1867, p. 218.
2. For recent accounts of James Barry's career, see M. Garber, *Vested Interests: Cross-Dressing and Cultural Anxiety*, Penguin, Harmondsworth, 1992, pp. 203–5; C. Blake, *The Charge of the Parasols: Women's Entry to the Medical Profession*, Women's Press, London, 1990, pp. 89–90.
3. 'Medical Women', *Australian Medical Journal*, 10, 1865, p. 235.
4. 'Dr Mary Walker and Dr Elizabeth Blackwell and Miss Garrett', *Victoria Magazine*, 8, 1867, pp. 232–3.
5. Elizabeth Garrett to Emily Davies, 5 September 1860, Autograph Letters Collection, Fawcett Library.
6. Elizabeth Garrett to Emily Davies, 7 August 1860.
7. Elizabeth Garrett to Emily Davies, 20 September 1860.
8. E. Blackwell, 'Letter to Young Ladies Desirous of Studying Medicine' (1860) reprinted in C. A. Lacey (ed.), *Barbara Leigh Smith Bodichon and the Langham Place Group*, p. 457.
9. E. Blackwell, 'Letter to Young Ladies', p. 458.
10. Elizabeth Blackwell to Barbara Bodichon, 30 December 1860, reprinted in T. P. Fleming, 'Dr Elizabeth Blackwell on Florence Nightingale', *Columbia University Columns*, 6, 1956, p. 43.

11. St Mary's Dispensary, *Annual Report*, 1867, p. 4. This Dispensary became the New Hospital for Women in 1872, later changing its name again to the Elizabeth Garrett Anderson Hospital.
12. R. R. Winter to Florence Nightingale, 17 October 1866, Nightingale Correspondence, GLRO, HI/ST/NC2/V/66.
13. Report to the Nightingale Fund Committee by Mrs Wardroper and Dr Whitfield, 1 August 1867, Nightingale Correspondence, GLRO, HI/ST/NC18/20/6/2.
14. Henry Bonham Carter to Florence Nightingale, 29 November 1867, Nightingale Papers, BL Add. MSS 47,715, f. 116.
15. Lucy Osburn to Florence Nightingale, 23 October 1866, Nightingale Correspondence, GLRO, H1/ST/NC2/V/6/66 [original emphasis].
16. Florence Nightingale, Comments on the New Hospital for Women, 1888, Elizabeth Garrett Anderson Hospital Records, GLRO, HI3/EGA/228/2 [original emphasis].
17. M. Scharlieb, *Reminiscences*, Williams & Norgate, London, 1924, p. 29; For further analysis of British medical women in India, see K. Jayawardena, *The White Woman's Other Burden: Western Women and South Asia During British Rule*, Routledge, New York and London, 1995, pp. 75–90; A. Burton, *Burdens of History: British Feminists, Indian Women, and Imperial Culture, 1865–1915*, University of North Carolina Press, Chapel Hill and London, 1994, pp. 112–24.
18. 'Victoria Discussion Society', *Victoria Magazine*, 14, 1870, p. 519.
19. Samuel Gregory, *Female Physicians*, reprinted from the *English Woman's Journal* by the Female Medical Society, n.d., p. 1.
20. 'The First English School of Medicine for Women', *Alexandra Magazine*, June 1865, p. 328.
21. Elizabeth Garrett to Emily Davies, 15 June 1860 [original emphasis].
22. 'Victoria Discussion Society', *Victoria Magazine*, 14, 1870, p. 521.
23. 'Lady Doctors', *Victoria Magazine*, 3, 1864, pp. 127–30.
24. 'Victoria Discussion Society', *Victoria Magazine*, 14, 1870, p. 511.
25. Quoted in M. Benn, *Predicaments of Love*, p. 130.
26. Nightingale Probationers' Record Book A, Nightingale Training School Records, GLRO, H1/ST/NTS/C4/1.
27. Evidence of Lucy Osburn, First Report, Royal Commission on Public Charities, New South Wales Legislative Assembly, *Votes and Proceedings*, vol. 6, 1873–74, p. 243 (q. 7551).
28. E. Blackwell, 'Medicine as a Profession for Women' (1862) in C. A. Lacey (ed.), *Barbara Leigh Smith Bodichon and the Langham Place Group*, p. 412.
29. Nightingale Probationers' Record Book A, Nightingale Training School Records, GLRO, H1/ST/NTS/C4/1.
30. *Australian Medical Gazette*, August 1870, p. 139.
31. R. Morantz-Sanchez, *Sympathy and Science: Women Physicians in American Medicine*, Oxford University Press, New York and London, 1985, pp. 185 ff; R. Morantz-Sanchez, 'Feminist Theory and Historical Practice: Rereading Elizabeth Blackwell' in A. Shapiro (ed.), *Feminists' Revision History*, Rutgers University Press, New Brunswick & New Jersey, 1994, pp. 95–119.

32. For commentary on this historiography, see B. Caine, *Victorian Feminists*, Oxford University Press, Oxford, 1992, pp. 2–17; L. Davidoff, 'Regarding Some "Old Husbands" Tales': Public and Private in Feminist History' in *Worlds Between: Historical Perspectives on Gender and Class*, Polity, Cambridge, 1995, pp. 227–76.
33. B. Caine, *Victorian Feminists*, pp. 16–17.
34. See S. Holton, 'Feminine Authority and Social Order', pp. 59–72.
35. Elizabeth Blackwell to Barbara Bodichon, 30 December 1860, reprinted in Fleming, 'Dr Elizabeth Blackwell on Florence Nightingale', p. 41.
36. Florence Nightingale to John Stuart Mill, 12 September 1860, BL Add. MSS 45, 787 ff. 11–12.
37. Florence Nightingale to Sir Harry Verney, n.d., copy in Claydon Letters, Contemporary Medical Archives, Wellcome Institute for the History of Medicine [original emphasis].
38. Florence Nightingale to Sir Harry Verney, 16 April 1867, Claydon Letters, [original emphasis].
39. Florence Nightingale to Henry Bonham Carter, 5 October 1867, BL Add. MSS 47,715, f. 74.
40. 'I could almost have thought that the important object of having fully qualified Female Physicians would rather have been brought about by beginning with a Lying-In Hospital & also a Female Hospital where women could be fully trained as Physician Accoucheuses & Physicians for the Diseases of Women & children at the bedside as well as by lectures.' Florence Nightingale, n.d., BL Add. MSS 45,804, ff. 218–19.
41. Florence Nightingale to Harry Verney, 16 April 1867, Claydon Letters.
42. S. Holton, 'Feminine Authority and Social Order', p. 69.
43. F. Nightingale, 'Nursing the Sick' in L. Seymer (ed.), *Selected Writings*, pp. 334–6; See also S. Holton, 'Feminine Authority and Social Order', pp. 60–61; C. E. Rosenberg, 'Florence Nightingale on Contagion', p. 126.
44. See L. Jordanova, *Sexual Visions*, pp. 19–42.
45. D. Haraway, *Primate Visions: Gender, Race and Nature in the World of Modern Science*, Verso, London and New York, 1992, p. 54.
46. See for example, C. Merchant, *Death of Nature: Women, Ecology and the Scientific Revolution*, HarperCollins, New York, 1990; V. Plumwood, *The Mastery of Nature*, Routledge, London and New York, 1993.
47. S. Gregory, *Female Physicians*, p. 3.
48. 'Female Physicians', *Social Science Review*, 21 June 1862, pp. 17–18. *Social Science Review* was edited by Richardson, and it is reasonable to presume him to be the author of this editorial.
49. B. W. Richardson, 'Woman as a Sanitary Reformer', p. 189 [original emphasis].
50. B. W. Richardson, 'Woman as a Sanitary Reformer', p. 199.
51. 'Medical Women', *Australian Medical Journal*, 10, 1865, pp. 234–35.
52. J. Comaroff, 'Medicine: Symbol and Ideology' in P. Wright and A. Treacher (eds), *The Problem of Medical Knowledge*, Edinburgh University Press, Edinburgh 1982, pp. 56–7.
53. C. E. Rosenberg, 'Florence Nightingale on Contagion'.
54. F. P. Cobbe, 'The Medical Profession and its Morality', *Modern Review*, 2, 1881, p. 309.

55. F. P. Cobbe, 'Hygeiolatry' in her *Peak of Darien*, Williams and Norgate, London, 1882, p. 86, p. 89.
56. For elaborations of the idea of the place of purity and morality in early feminism, see M. Vicinus, *Independent Women*, pp. 17–18; S. Jeffreys, *The Spinster and her Enemies: Feminism and Sexuality, 1880–1930*, Pandora, London and Boston, 1985; L. Bland, *Banishing the Beast: English Feminism and Sexual Morality, 1885–1914*, Penguin, Harmondsworth, pp. 95–123.
57. E. Blackwell, 'The Influence of Women in the Profession of Medicine' (1889) in her *Essays in Medical Sociology*, vol. 2, Arno Press, New York, 1972, pp. 9–12.
58. E. Blackwell, 'The Influence of Women', p. 6.
59. E. Blackwell, *Christianity in Medicine*, J.F. Nock, London, 1890, p. 2.
60. Blackwell, 'The Influence of Women', pp. 28–9.
61. Elizabeth Blackwell to Florence Nightingale, 25 July no year, BL Add. MSS 45,802, f. 237; See also, E. Blackwell, *Pioneer Work for Women*, J. M. Dent, London, 1914, p. 199.
62. M. Scharlieb, *Reminiscences*, p. 208.
63. L. Martindale, *The Woman Doctor and her Future*, Mills & Boon, London, 1922, p. 133. For further discussion of women doctors' involvement in feminism and eugenics in the early twentieth century, see A. Bashford, 'Edwardian Feminists and the Venereal Disease Debate in England' in B. Caine (ed.), *The Woman Question in England and Australia*, University Printing Service, University of Sydney, 1994, pp. 58–85; L. Bland, *Banishing the Beast*, pp. 222–49.

6 Dissecting the Feminine: Women Doctors and Dead Bodies in the Late Nineteenth Century

1. L. Jordanova, *Sexual Visions*.
2. J. Kristeva, *Powers of Horror: An Essay on Abjection*, Columbia University Press, New York, 1982.
3. R. Richardson, *Death, Dissection and the Destitute*, Routledge, London and New York, 1987.
4. R. Richardson, *Death, Dissection and the Destitute*, p. 279.
5. R. Richardson, *Death, Dissection and the Destitute*, p. 76.
6. *University of Sydney Calendar*, Sydney, 1890, p. 176.
7. For the development of microscopy and bacteriology, see W. F. Bynam, *Science and the Practice of Medicine in the Nineteenth Century*, Cambridge University Press, Cambridge, 1994, pp. 123–32.
8. J. Maclise, *Surgical Anatomy*, John Churchill, London, 1851, preface, n.p.
9. 'A Night in the Dissecting Room,' *Speculum*, 18, 1889, pp. 47–8.
10. J. Bland-Sutton, *The Story of a Surgeon*, Methuen, London, 1930, p. 39.
11. Elizabeth Garrett to Emily Davies, 23 March 1861, Autograph Letters Collection, Fawcett Library.
12. 'Topic of the Day,' *Medical Times and Gazette,* 26 November 1870, p. 622.

13. Female Medical Students to the Faculty of Medicine, 27 September 1887, Admission of Women Correspondence, University of Melbourne Archives.
14. Female Medical Students to the Chancellor, Vice- Chancellor and Members of the Council, University of Melbourne, 22 April 1895, Admission of Women Correspondence; See also Clara Stone to Women Medical Students Society, 9 September 1907, Women Medical Students Society Correspondence, University of Melbourne Archives.
15. P. Ariès, *The Hour of our Death*, Alfred A. Knopf, New York, 1981, p. 370.
16. Pisanus Fraxi, *Index Librorum Prohibitorum*, privately printed, London, 1877, p. 415. Noted in R. Richardson, *Death, Dissection and the Destitute*, p. 95.
17. L. Jordanova, *Sexual Visions*, p. 1. Cited also in E. Showalter, *Sexual Anarchy: Gender and Culture at the Fin de Siècle*, Virago, London, 1992, p. 131.
18. L. Jordanova, *Sexual Visions*, pp. 87–110; See also E. Showalter, *Sexual Anarchy*, pp. 127–43.
19. E. Bronfen, *Over Her Dead Body: Death, Femininity and the Aesthetic*, Manchester University Press, Manchester, 1992.
20. See for example J. Maclise, *Surgical Anatomy*; C. Heath, *Practical Anatomy: A Manual of Dissections*, J. & A. Churchill, London, 1874; J. Cleland, *A Directory for the Dissection of the Human Body*, Smith Elder & Co., London, 1888.
21. 'Topic of the Day', *Medical Times and Gazette*, 26 November 1870, p. 622.
22. *Medical Times and Gazette*, 26 November 1870, p. 628.
23. E. Bronfen, *Over Her Dead Body*, p. 102
24. 'Lady Students', *Speculum*, 31, 1895, p. 23.
25. D.J. Cunningham, *Manual of Practical Anatomy*, vol. 1, Young J. Pentland, Edinburgh and London, 1893, p. 348.
26. Elizabeth Garret to Emily Davies, 23 March 1861.
27. Alice Newton-Tabrett to editor, *Sydney Morning Herald*, 12 February 1906, p. 10.
28. E. Grosz, *Volatile Bodies*, p. 203.
29. *Age*, 22 January 1887.
30. R. Pringle and S. Collings, 'Women and Butchery: Some Cultural Taboos', *Australian Feminist Studies*, 17, 1993, pp. 29–45.
31. 'Lady Doctors', *Victoria Magazine*, 3, 1864, p. 130.
32. Victoria Discussion Society, *Victoria Magazine*, 14, 1870, p. 509.
33. Spencer Wells, 'Castration in Mental and Nervous Diseases', *American Journal of the Medical Sciences*, 91-2, 1886, pp. 470–1. Cited in E. Showalter, *Sexual Anarchy*, pp. 133–4.
34. R. Pringle and S. Collings, 'Women and Butchery', p. 39.
35. See M. A. Elston, 'Women and Anti-Vivisection in Victorian England 1870–1900' in N. Rupke (ed.) *Vivisection in Historical Perspective*, Croom Helm, London, 1987, pp. 279–81.
36. See C. Lansbury, 'Gynaecology, Pornography, and the Antivivisection Movement', *Victorian Studies*, 28, 1985, p. 415.

37. See J. Walkowitz, *City of Dreadful Delight: Narratives of Sexual Danger in Late Victorian London*, University of Chicago Press, Chicago, 1992, pp. 210, 224.
38. See S. K. Kent, *Sex and Suffrage in Britain*, p. 119; J. Walkowitz, *City of Dreadful Delight*, pp. 92–3.
39. E. Bronfen, *Over Her Dead Body*, 87.
40. In both miasmatic and 'germ theory' understandings of hospital fevers, corpses and the postmortem theatres were considered dangerous and polluting. When outbreaks of fevers did occur, the architecture and position of the postmortem theatre and the cleanliness of medical students and others in contact with corpses was often immediately investigated. One example is the closure of the King's College Hospital maternity ward in the late 1860s due to epidemic of puerperal fever. See King's College Hospital Committee of Management Minutes, 8 January 1868, King's College Archives, London.
41. E. Bronfen, *Over Her Dead Body*, p. 88.
42. M. Foucault, *Birth of the Clinic*, Vintage, New York, 1975, p. 145.
43. M. Douglas, *Purity and Danger*, p. 98.
44. J. Bland-Sutton, *The Story of a Surgeon*, pp. 38–9.
45. J. Kristeva, *Powers of Horror*, p. 4, p. 9.
46. Cited in R. Kramer, *Maria Montessori: A Biography*, Blackwell, Oxford, 1978, p. 42.
47. J. Kristeva, *Powers of Horror*, p. 3 [original emphasis].
48. E. Grosz, *Sexual Subversions*, Allen & Unwin, Sydney, 1989, p. 75.
49. R. Kramer, *Maria Montessori*, p. 42.
50. E. Bronfen, *Over Her Dead Body*, p. 255.
51. E. Bronfen, *Over her Dead Body*, p. xii.
52. E. Bronfen, *Over Her Dead Body*, p. 86.
53. L. Jordanova, *Sexual Visions*, p. 150.

7 Sterile Bodies: Germs and the Gendered Practitioner

1. A. J. Youngson, *The Scientific Revolution in Victorian Medicine*, Croom Helm, London, 1979, p. 23. See also the work of Frederick F. Cartwright, who writes that what 'produce[d] a revolution in medical thinking and practice were the "cell theory" and the "germ theory"....[which] must be accorded the first place in changing medicine from an empirical art into a science'. F. F. Cartwright, *A Social History of Medicine*, Longman, London and New York, 1977, pp. 149–50.
2. M. Douglas, *Purity and Danger*, p. 36; See also G. Vigarello, *Concepts of Cleanliness*, pp. 202–14; D. Lupton, *The Imperative of Health: Public Health and the Regulated Body*, Sage, London, 1995, pp. 36–7.
3. W. F. Bynam, *Science and the Practice of Medicine in the Nineteenth Century*, p. 132; P.S. Hardy, ' "Surgical Spirit": Listerism and the Medical Profession In New South Wales, 1867–1889', PhD thesis, University of New South Wales, 1990, p. 2.

4. F. F. Cartwright, 'Antiseptic Surgery', pp. 90–1; F. F. Cartwright, *The Development of Modern Surgery*, Arthur Barker, London, 1967, p. 56, p. 78.
5. T. H. Pennington, 'Listerism, its Decline and its Persistence: The introduction of aseptic surgical techniques in three British teaching hospitals, 1890–99', *Medical History*, 39, 1995, pp. 35–60.
6. C. McBurney, 'The Technic of Aseptic Surgery' in A. P. Gould and J. C.Warren (eds), *The International Text Book of Surgery*, London and Philadelphia, WB Saunders, 1902, p. 270.
7. A. Thomson and A. Miles, *A Manual of Surgery*, Young J. Pentland, Edinburgh and London, 1906, p. 10 [original emphasis].
8. For the increasing list of items deemed possible and necessary to sterilise, see C. B. Lockwood, 'Further Report on Aseptic and Septic Surgical Cases, with special reference to the disinfection of materials and the skin', *BMJ*, II, 1896, pp. 59–62; C. B. Lockwood, 'An Address on the Organisation of Aseptic Operations and Some of the Causes of Failure', *BMJ*, I, 1900, pp. 429–32.
9. For new developments towards what has come to be known as 'scientific medicine' in the late nineteenth century, see R. E. Kohler, 'Medical Reform and Biomedical Science – a Case Study', and R. C. Maulitz, ' "Physician versus Bacteriologist": The Ideology of Science in Clinical Medicine', in M. J. Vogel and C. E. Rosenberg (eds), *The Therapeutic Revolution*, University of Pennsylvania Press, Philadelphia, 1979, pp. 27–66, 91–107.
10. J. C. Da Costa, *Modern Surgery: General and Operative*, W.B. Saunders, London, 1900, p. 17.
11. A. Thomson and A. Miles, *A Manual of Surgery*, p. 10.
12. W. Playfair, *A Treatise on the Science and Practice of Midwifery*, Smith Elder & Co., London, 9th edn, 1898, pp. 365, 374.
13. P. S. Hardy, 'Surgical Spirit', p. 16, pp. 19–20, p. 289.
14. For example: 'What is known as the antiseptic method we owe to the splendid labors of Lord Lister, and the aseptic method is but a natural evolution of the antiseptic method.' Da Costa, *Modern Surgery*, p. 24.
15. C. Lawrence and R. Dixey, 'Practising on Principle: Joseph Lister and the Germ Theory of Disease', in C. Lawrence (ed.), *Medical Theory, Surgical Practice*, Routledge, London and New York, 1992, p. 154.
16. C. Lawrence and R. Dixey, 'Practising on Principle', p. 153; See also L. Granshaw, ' "Upon this principle I have based a practice": the development and reception of antisepsis in Britain, 1867–1890', in J. Pickstone (ed.), *Medical Innovations in Historical Perspective*, Macmillan, London, 1992.
17. S. Holton, 'Feminine Authority and Social Order', pp. 70–1.
18. See for example, E. Lückes, *Lectures on General Nursing*; C. Wood, *A Handbook of Nursing for Hospital and Home*, Cassell & Co., London and Melbourne, n.d. (189?); E. C. Laurence, *Modern Nursing in Hospital and Home*, Scientific Press, London, 1907.
19. C. E. Rosenberg, 'The Therapeutic Revolution: Medicine, Meaning, and Social Change in Nineteenth-Century America', *Perspectives in Biology and Medicine*, 20, 1977, pp. 485–506.
20. C. E. Rosenberg, 'The Therapeutic Revolution', pp. 488–9, p. 498.

21. C. E. Rosenberg, 'The Therapeutic Revolution', p. 490.
22. E. Lückes, *Lectures on General Nursing*, p. 101.
23. G. B. Burbidge, *Lectures for Nurses*, Australasian Medical Publishing, Glebe, 1935, p. 63; See also E. C. Laurence, *Modern Nursing*, pp. 67-9.
24. C. E. Rosenberg, 'The Therapeutic Revolution', p. 489.
25. E. Lückes, *Lectures on General Nursing*, pp. 161-62, p. 165.
26. E. Lückes, *Lectures on General Nursing*, p. 162.
27. A. Munro, *The Science and Art of Nursing the Sick*, p. 103 [original emphasis].
28. C. Wood, *A Handbook of Nursing*, p. 118.
29. R. Strong, 'Hygiene of the Sick-Room' in H. Morten (ed.), *A Complete System of Nursing written by Medical Men and Nurses*, Sampson, Low, Marston & Co., London, 1903, p. 4.
30. F. G. Holden, 'Diphtheria', *Humanity and Health*, January 1894, pp. 13-14 [original emphasis].
31. C. E. Rosenberg, 'The Therapeutic Revolution', p. 489.
32. E. Lückes, *Lectures on General Nursing*, pp. 197-200.
33. P. S. Hardy, 'Surgical Spirit', p. 21.
34. C. Wood, *A Handbook of Nursing*, pp. 62-3 [original emphasis].
35. E. Lückes, *Lectures on General Nursing*, p. 221.
36. C. Wood, *A Handbook of Nursing*, p. 151.
37. Louise Creighton, speech at the Annual Meeting of New Hospital, London, reported in the *Philanthropist*, March 1897, Newscutting Album 1881-1917, records of the Elizabeth Garrett Anderson Hospital, GLRO, H13/EGA/144.
38. A. Miles, *Surgical Ward Work and Nursing*, Scientific Press, London, 1899.
39. I. Stewart and H. Cuff, *Practical Nursing*, William Blackwood, Edinburgh and London, 1903, p. 350.
40. H. W. G. Macleod, *Hygiene for Nurses*, Smith Elder, London, 1911.
41. See for example, A. Munro, *The Science and Art of Nursing the Sick*; E. J. Domville, *A Manual for Hospital Nurses*, J. & A. Churchill, London, 1878; H. Morten (ed.), *A Complete System for Nursing;* W. J. Hadley, *Nursing: General, Medical and Surgical*, J. & A. Churchill, London, 1902.
42. H. W. G. Macleod, *Hygiene*, pp. 154-7.
43. See *The Lancet*, II, 1900, p. 1668; *BMJ*, I, 1898, p. 1355; *BMJ*, II, 1902, p. 1819.
44. 'Antiseptics in the Larder', *BMJ*, I, 1895, p. 907.
45. A. C. Abbott, *The Principles of Bacteriology*, H. K. Lewis, London, 1902.
46. For discussion of 'methods of purification', meaning sterilisation, see F. F. Burchard, 'A Discussion on the Present Position of the Aseptic Treatment of Wounds', *BMJ*, II, 1904, p. 795.
47. F.F. Burchard, 'A Discussion on the Present Position of the Aseptic Treatment of Wounds', *BMJ*, II, 1904, p. 797
48. N. J. Fox, 'Scientific Theory Choice and Social Structure: The Case of Joseph Lister's Antisepsis, Humoral Theory and Asepsis', *History of Science*, 26, 1988, p. 391.
49. N. J. Fox, 'Scientific Theory Choice', p. 390.

50. C. Leedham-Green, 'A Bacteriological Inquiry into the Relative Value of Various Agents Used in the Disinfection of the Hands', *BMJ*, II, 1896, p. 1109.
51. C. Yelverton Pearson, 'Observations on Sterilization of the Hands', *BMJ*, II, 1905, p. 785 [original emphasis]. For other examples of such methods and experiments, see C. B. Lockwood, 'Further Report on Aseptic and Septic Surgical Cases', *BMJ*, II, 1896, pp. 59–62; J. R. Collins, 'Bacteriological Inquiry into the Sterilization of Hands', *BMJ*, I, 1904, pp. 1364–66.
52. 'The Cleansing of the Hands', *BMJ*, I, 1901, p. 912.
53. R. Howard, *The House Surgeon's Vade Mecum*, Edward Arnold, London, 1911, p. 5.
54. J. R. Collins, 'An Experimental Inquiry into the Infection of Operative Wounds, from the skin, the breath, and the air', *BMJ*, II, 1905, pp. 121–5.
55. 'An Aseptic Operating Cap', *BMJ*, II, 1904, p. 22.
56. 'The Veiled Operator', *BMJ*, I, 1899, p. 1296.
57. R. Howard, *The House Surgeon's Vade Mecum*, p. 5.
58. See for example, R. G. Richardson, *The Surgeon's Tale*, Charles Scribners, New York, 1958, pp. 32–37; D. de Moulin, *A History of Surgery*, Martinus Nijhoff, Dordrecht, 1988, pp. 294–95; F. F. Cartwright, 'Antiseptic Surgery', p. 88.
59. See Dr Snow Beck, 'On a Case of Puerperal Fever', *The Lancet*, II, 1867, pp. 805–6; F. Churchill, *On the Theory and Practice of Midwifery*, p. 676; K. Codell Carter, 'Ignaz Semmelweis, Carl Mayrhofer, and the Rise of Germ Theory', p. 46.
60. For example, there is no mention at all of puerperal fever and obstetric practice in the following two articles: Lord Lister, 'Early Researches Leading up to the Antiseptic System of Surgery', *The Lancet*, II, 1900, pp. 985–93; 'Lord Lister and Antiseptic Surgery: The History of a Revolution', *BMJ*, II, 1902, pp. 1841–8;
61. J. Randers-Pehrson, *The Surgeon's Glove*, Garles C. Thomas, Springfield, 1960, pp. 2–3.
62. K. W. Monarrat, *Surgical Technics in Hospital Practice*, John Wright & Co., Bristol, 1898, pp. 7–10.
63. See for example F. Churchill, *On the Theory and Practice of Midwifery*, p. 691
64. C. McBurney, 'The Technic of Aseptic Surgery', p. 288.
65. See for example, R. Howard, *The House-Surgeon's Vade Mecum*, pp. 1–2.
66. For example, Da Costa does not mention a patient, a living body or a human agent throughout his first chapter on bacteriology. See also J.C. Warren, *Surgical Pathology and Therapeutics*, W. B. Saunders, Philadelphia, chapter one; W. Lusk, *The Science and Art of Midwifery*, HK Lewis, London, 1882, pp. 616–18.
67. F. Nightingale, 'Nurses, Training of' in L. Seymer (ed.) *Selected Writings*, p. 351.
68. F. Treves, 'The Surgeon in the Nineteenth Century', *The Lancet*, II, 1900, p. 316.

Bibliography

PRIMARY MATERIAL

Aveling, James, *English Midwives, their History and Prospects* (1872) Hugh K. Elliott, London, 1967.
Barnes, Robert, *On Antiseptic Midwifery and Septicaemia in Midwifery*, W. M. Wood, New York, 1882.
Bell, Joseph, *Notes on Surgery for Nurses*, Oliver & Boyd, Edinburgh, 1899.
Blackwell, Elizabeth, *Pioneer Work for Women*, J. M. Dent, London, 1914.
—— *Essays in Medical Sociology* (1902) Arno Press, New York, 1972.
—— *Scientific Method in Biology*, Elliot Stock, London, 1898.
—— *Christianity in Medicine*, J. F. Nock, London, 1890.
Blyth, Alexander, *A Dictionary of Hygiene and Public Health*, Charles Griffith, London, 1876.
Chadwick, Edwin, 'Address on Public Health', *Transactions of the National Association for the Promotion of Social Science*, 1860, pp. 574–606.
—— *Report on the Sanitary Condition of the Labouring Population of Great Britain* (1842), edited with an introduction by M. W. Finn, Edinburgh University Press, Edinburgh, 1965.
Churchill, Fleetwood, *A Manual for Midwives and Monthly Nurses*, Longman, London, 1856.
Cleland, John, *A Directory for the Dissection of the Human Body*, Smith Elder & Co., London, 1888.
Cobbe, Frances Power, 'The Medical Profession and its Morality', *Modern Review*, 2, 1881, pp. 296–328.
Coffin, A.I., *Treatise on Midwifery*, A. I. Coffin, London, 1866.
Cowper, W., 'Address on Public Health', *Transactions of the National Association for the Promotion of Social Science*, 1859, pp. 106–21.
Cunningham, D. J., *Manual of Practical Anatomy*, vol. 1, Young J. Pentland, Edinburgh and London, 1893.
Davies, Emily, letters in Autograph collection of letters, Fawcett Library.
Domville, Edward James, *A Manual for Hospital Nurses and Others Engaged in Attending the Sick*, Churchill, London, 1878.
'Dr Mary Walker and Dr Elizabeth Blackwell and Miss Garrett', *Victoria Magazine*, 8, 1867, pp. 232–3.
Druitt, Robert, *Surgeon's Vade Mecum: A Manual of Modern Surgery*, Henry Renshaw & John Churchill, London, 1865.
Earl of Shaftesbury, 'Address on Public Health', *Transactions of the National Association for the Promotion of Social Science*, 1858, pp. 84–95.
Edmunds, James, *Inaugural Address delivered for the Female Medical Society*, 3 October 1864, London, 1864.
—— 'The Introductory Address Delivered for the Second Session of the Female Medical Society', *Victoria Magazine*, 6, November 1865, pp. 47–55.

Bibliography

—— 'Female Medical Society and Morality in Childbirth', *Medical Times and Gazette*, 25 August 1866, p. 206.

—— *The Introductory Address Delivered for the Female Medical Society*, n.d. bound in Women's Medical Education Pamphlets, Fawcett Library.

Fawcett, Millicent Garrett, 'The Medical and General Education of Women', *Fortnightly Review*, 4, 1868, pp. 554–71.

Galton, Douglas, *An Address on the General Principles which Should be Observed in the Construction of Hospitals*, Macmillan & Co., London, 1869.

Garrett, Elizabeth, 'Hospital Nursing', *Transactions of the National Association for the Promotion of Social Sciences*, 1866, pp. 472–78.

—— 'Public Hospitals and Dispensaries', *Transactions of the National Association for the Promotion of Social Science*, 1868, pp. 464–7.

Garrett Anderson, Elizabeth, letters in Autograph Collection of Letters, Fawcett Library.

Gregory, Samuel, *Female Physicians*, Female Medical Society, London, n.d.

Haddon, Charlotte, 'Nursing as a Profession for Ladies', *St Paul's Magazine*, August 1871, pp. 458–61.

Holland, Sydney, *A Talk to the Nurses of the London Hospital*, Whitehead, Morris, London, c. 1897.

Hornsby, John Allan and Schmidt, Richard E., *The Modern Hospital: Its Inspiration, its Architecture, its Equipment, its Operation*, W. B. Saunders, Philadelphia and London, 1913.

'Hospital Nurses as the Are and as they Ought to Be', *Frasers Magazine*, 37, 1848, pp. 539–42.

Howard, Russell, *The House Surgeon's Vade Mecum*, Edward Arnold, London, 1911.

Jameson, Anna, *Sisters of Charity, Catholic and Protestant, Abroad and at Home*, Longman, Brown, Green & Longmans, London, 1855.

Laurence, Eleanor Constance, *Modern Nursing in Hospital and Home*, Scientific Press, London, 1907.

Loane, M. and Bowers, H., *The District Nurse as Health Missioner*, Women's Printing, London, n.d. [c. 1908].

Lückes, Eva, *Lectures on General Nursing*, Kegan, Paul, Trench & Co, London, 1884.

—— *Hospital Sisters and their Duties*, Scientific Press, London, 1893.

Mackintosh, Donald J., *The Construction, Equipment and Management of a General Hospital*, William Hodge & Co., Edinburgh and Glasgow, 1909.

Macleod, Herbert W. G., *Hygiene for Nurses: Theoretical and Practical*, Smith Elder, London, 1911.

Maclise, Joseph, *Surgical Anatomy*, John Churchill, London, 1851.

Mann, R. J., *Domestic Economy and Household Science*, Edward Stanford, London, 1878.

Martindale, Louisa, *The Woman Doctor and Her Future*, Mills & Boon, London, 1922.

Memorials of Agnes Elizabeth Jones, Strahan & Co., London, 1871.

Miles, Alexander, *Surgical Ward Work and Nursing: A Handbook for Nurses and Others*, Scientific Press, London, 1899.

Monarrat, K. W., *Surgical Technics in Hospital Practice*, John Wright & Co., Bristol, 1898.

Morten, Honnor (ed.), *A Complete System of Nursing written by Medical Men and Nurses*, Sampson, Low, Marston & Co., London, 1903.
Mouat, Frederic J. and Snell, H. Saxon, *Hospital Construction and Management*, J. & A. Churchill, London, 1883.
Munro, Aeneas, *The Science and Art of Nursing the Sick*, James Maclehouse, Glasgow, 1873.
Murphy, E. W., *Lectures on the Principles and Practice of Midwifery*, Walton & Maberly, London, 1862.
Nightingale, Florence, 'Notes on the Sanitary Conditions of Hospitals and on Defects in the Construction of Hospital Wards', *Transactions of the National Association for the Promotion of Social Science*, 1858, pp. 462–82.
—— 'Sick Nursing and Health Nursing' in Isabel A. Hampton (ed.), *Nursing of the Sick*, 1893, MacGraw-Hill, 1949.
—— *Selected Writings of Florence Nightingale* (Lucy Seymer ed.), Macmillan, London, 1954.
—— *Notes on Nursing, What it is and what it is not* (1859), Dover Publications, New York 1969.
—— *Introductory Notes on Lying-In Institutions*, Longmans, Green and Co., London, 1871
Nightingale Correspondence, Claydon Letters, Contemporary Medical Archives Centre, Wellcome Institute for the History of Medicine, London.
Nightingale Correspondence, British Library (BL).
Nightingale Collection, Greater London Record Office.
'Nurses Wanted', *Cornhill Magazine*, 11, 1865, pp. 409–25.
Oppert, F., *Hospitals, Infirmaries and Dispensaries: Their Construction, Interior Arrangement and Management*, John Churchill, London, 1867.
Parkes, Bessie Rayner, 'The Ladies' Sanitary Association', *English Woman's Journal*, 3, 1859, pp. 73–85.
—— 'At a Nurses' Training School', *Alexandra Magazine*, February 1865, pp. 65–71.
Pickford, James H., *Hygiene or Health*, John Churchill, London, 1858.
Playfair, William, *A Treatise on the Science and Practice of Midwifery*, Smith Elder & Co., London, ninth edn, 1898.
Powers, Susan, 'The Details of Woman's Work in Sanitary Reform', *English Woman's Journal*, 3, 1859, pp. 217–27, 316–24.
—— *Remarks on Woman's Work in Sanitary Reform*, Ladies' Sanitary Association, London, n.d.
Richardson, Benjamin Ward, 'Female Physicians', *Social Science Review*, June 1862, pp. 17–18.
—— *Hygeia: A City of Health*, London, 1875.
—— 'Woman as a Sanitary Reformer', *Transactions of the Sanitary Institute*, 2, 1880, pp. 183–202.
Scharlieb, Mary, *Reminiscences*, Williams & Norgate, London, 1924.
Simon, John, *Public Health Reports*, 2 vols, J.&A. Churchill, London, 1887.
Smith, William Robert, *Lectures on Nursing*, J.&A. Churchill, London, 1875.
Stanley, Mary, *Hospitals and Sisterhoods*, John Murray, London, 1855.
Steele, J. C. 'Nursing and Nursing Institutes', *Sanitary Record*, January 1875, pp. 4–6.

Strangford, Lady, *Hospital Training for Ladies: An Appeal to the Hospital Boards in England*, Harrison & Sons, London, 1874.
Talley, W., *He, or Man-Midwifery and the Results: or, Medical Men in the Criminal Courts*, Job Caudwell, London, 1863.
Thomson, Alexander, *Lecture on Sanitary Reform*, George and Robert King, Aberdeen, 1860.
Thomson, Anthony Todd, *Domestic Management of the Sick- Room*, Longman, Brown, Green & Longmans, London, 1845.
Trench, M., 'Sick Nurses', *Macmillans Magazine*, 34, 1876, pp. 422–9.
Twining, Louisa, *Nurses for the Sick with a Letter to Young Women*, Longmans and Roberts, London, 1861.
Watson, J.K., *A Handbook for Nurses*, Scientific Press, London, 1899.
West, Charles, *Medical Women: A Statement and an Argument*, J.&A. Churchill, London, 1878.
Winn, James, *Outlines of Midwifery*, Longmans, Brown, Green & Longman, London, 1854.
Wood, Catherine, *A Handbook of Nursing for Hospital and Home*, Cassell & Co., London and Melbourne, n.d. (189–).

SECONDARY MATERIAL

Abel-Smith, Brian, *A History of the Nursing Profession*, Heinemann, London, 1960.
—— *The Hospitals 1800–1948*, Heinemann, London, 1964.
Ackerknecht, E.H., 'Anticontagionism between 1821 and 1867', *Bulletin of the History of Medicine*, 22, 1948, pp. 562–93.
Anderson, Amanda, *Tainted Souls and Painted Faces: The Rhetoric of Fallenness in Victorian Culture*, Cornell University Press, Ithaca and London, 1993.
Ariés, Philippe (trans. Helen Weaver), *The Hour of our Death*, Alfred A. Knopf, New York, 1981.
Armstrong, David, *Political Anatomy of the Body: Medical Knowledge in Britain in the Twentieth Century*, Cambridge University Press, Cambridge, 1983.
—— 'Public Health Spaces and the Fabrication of Identity', *Sociology*, 27, 1993, pp. 393–410.
Baly, Monica, *Florence Nightingale and the Nursing Legacy*, Croom Helm, London, 1986.
Bashford, Alison, 'Edwardian Feminists and the Venereal Disease Debate in England' in Barbara Caine (ed.), *The Woman Question in England and Australia*, University Printing Service, University of Sydney, 1994, pp. 58–85.
—— 'Nursing Bodies: The Gendered Politics of Health in Australia and England, c. 1860–1910', PhD thesis, University of Sydney, 1994.
—— 'Separatist Health: Changing Meanings of Women's Hospitals, c. 1870–1920' in Lilian R. Furst (ed.), *Women Healers and Physicians: Climbing a Long Hill*, University of Kentucky Press, Lexington, 1997.

Bibliography

Bell, Shannon, *Reading, Writing and Rewriting the Prostitute Body*, Indiana University Press, Bloomington and Indianapolis, 1994.
Benjamin, Marina (ed.), *Science and Sensibility: Gender and Scientific Enquiry 1780-1945*, Basil Blackwell, Oxford, 1991.
Benn, Miriam J., *Predicaments of Love*, Pluto Press, London, 1992.
Blake, Catriona, *The Charge of the Parasols: Women's Entry to the Medical Profession*, Women's Press, London, 1990.
Bland, Lucy, *Banishing the Beast: English Feminism and Sexual Morality, 1885-1914*, Penguin, Harmondsworth.
Bordo, Susan, 'Feminism, Foucault and the Politics of the Body' in Caroline Ramazanoglu (ed.), *Up Against Foucault: Explorations of Some Tensions between Foucault and Feminism*, Routledge, London and New York, 1993.
Bronfen, Elisabeth, *Over Her Dead Body: Death, Femininity and the Aesthetic*, Manchester University Press, Manchester, 1992.
Bynam, W. F., *Science and the Practice of Medicine in the Nineteenth Century*, Cambridge University Press, Cambridge, 1994.
Caine, Barbara, *Victorian Feminists*, Oxford University Press, Oxford, 1992.
——, E. A. Grosz and Marie de Lepervanche (eds), *Crossing Boundaries: Feminism and the Critique of Knowledges*, Allen & Unwin, Sydney, 1988.
Carter, K. Codell, 'Ignaz Semmelweis, Carl Mayrhofer, and the Rise of Germ Theory', *Medical History*, 29, 1985, pp. 33-53.
—— 'The Development of Pasteur's Concept of Disease Causation and the Emergence of Specific Causes in Nineteenth- Century Medicine', *Bulletin of the History of Medicine*, 65, 1991, pp. 528-48.
Cartwright, Frederick F., 'Antiseptic Surgery' in F. N. L. Poynter (ed.), *Medicine and Science in the 1860s*, Wellcome Institute for the History of Medicine, London, 1968.
Comaroff, Jean, 'Medicine: Symbol and Ideology' in Peter Wright and Andrew Treacher (eds), *The Problem of Medical Knowledge: Examining the Social Construction of Medicine*, Edinburgh University Press, Edinburgh, 1982.
Cooter, Roger, 'Anticontagionism and History's Medical Record' in Peter Wright and Andrew Treacher (eds), *The Problem of Medical Knowledge: Examining the Social Construction of Medicine*, Edinburgh University Press, Edinburgh, 1982.
Davidoff, Leonore, *Worlds Between: Historical Perspectives on Gender and Class*, Polity, Cambridge, 1995.
—— and Catherine Hall, *Family Fortunes: Men and Women of the English Middle-Class, 1780-1850*, Hutchinson, London and Melbourne, 1987.
Davies, Celia (ed.), *Rewriting Nursing History*, Croom Helm, London, 1980.
Dean, Mitchell, *The Constitution of Poverty: Towards a Genealogy of Liberal Governance*, Routledge, London and New York, 1991.
Donnison, Jean, *Midwives and Medical Men: A History of the Struggle for the Control of Childbirth*, Heinemann, London, 1977.
Douglas, Mary, *Purity and Danger: An Analysis of the Concepts of Pollution and Taboo* (1966), Routledge, London and New York, 1994.
Elston, Mary Ann, 'Women and Anti-Vivisection in Victorian England 1870-1900' in Nicolaas Rupke (ed.), *Vivisection in Historical Perspective*, Croom Helm, London, 1987.

Featherstone, M., Hepworth, M. and Turner, B. (eds), *The Body: Social Process and Cultural Theory*, Sage, London, 1991.
Fee, Elizabeth and Porter, Dorothy, 'Public Health, Preventive Medicine and Professionalization: England and America in the Nineteenth Century' in Andrew Wear (ed.), *Medicine in Society: Historical Essays*, Cambridge University Press, Cambridge, 1992.
Foucault, Michel, *The History of Sexuality: An Introduction*, Penguin, Harmondsworth, 1987.
—— 'The Politics of Health in the Eighteenth Century' in P. Rabinow (ed.), *The Foucault Reader*, Penguin, Harmondsworth, 1991.
—— *Discipline and Punish*, Penguin, Harmondsworth, 1991.
Fox, Nicholas J., 'Scientific Theory Choice and Social Structure: The Case of Joseph Lister's Antisepsis, Humoral Theory and Asepsis', *History of Science*, 26, 1988, pp. 367–97.
French, Roger and Andrew Wear (eds), *British Medicine in an Age of Reform*, Routledge, London and New York, 1991.
Gallagher, Catherine and Thomas Laqueur (eds), *The Making of the Modern Body*, University of California Press, Berkeley, 1987.
Gatens, Moira, *Imaginary Bodies: Ethics, Power and Corporeality*, Routledge, London and New York, 1996.
Grosz, Elizabeth, *Volatile Bodies: Toward a Corporeal Feminism*, Allen & Unwin, Sydney, 1994.
Hamlin, Christopher, 'Predisposing Causes and Public Health in Early Nineteenth-Century Medical Thought', *Social History of Medicine*, 5, 1992, pp. 43–70.
Helmstadter, Carol, 'Robert Bentley Todd, Saint John's House, and the Origins of the Modern Trained Nurse', *Bulletin of the History of Medicine*, 67, 1993, pp. 282–319.
Holton, Sandra, 'Feminine Authority and Social Order: Florence Nightingale's Conception of Nursing and Health Care', *Social Analysis*, 15, 1984, pp. 59–72.
Jacobus, Mary, Keller, Evelyn Fox and Shuttleworth, Sally (eds), *Body/Politics: Women and the Discourses of Science*, Routledge, London and New York, 1990.
Jordanova, Ludmilla, *Sexual Visions: Images of Gender in Science and Medicine between the Eighteenth and the Twentieth Centuries*, Harvester Wheatsheaf, 1989.
Kent, Susan Kingsley, *Sex and Suffrage in Britain*, Princeton University Press, Princeton, 1987.
Kohler, Robert E., 'Medical Reform and Biomedical Science – A Case Study', in Morris J. Vogel and Charles E. Rosenberg, (eds), *The Therapeutic Revolution*, University of Pennsylvania Press, Philadelphia, 1979.
Kristeva, Julia, *Powers of Horror: An Essay on Abjection*, Columbia University Press, New York, 1982.
Lacey, Candida Ann, (ed.), *Barbara Leigh Smith Bodichon and the Langham Place Group*, Routledge & Kegan Paul, New York and London, 1987.
Lansbury, Coral, 'Gynaecology, Pornography and the Anti-Vivisection Movement', *Victorian Studies*, 28, 1985, pp. 415–37.

Lawrence, Christopher, 'Sanitary Reformers and the Medical Profession in Victorian England' in Teizo Ogawa (ed.), *Public Health: Proceedings of the Fifth International Symposium on the Comparative History of Medicine East and West*, Saikon Publishing, Tokyo, 1981.
—— 'Incommunicable Knowledge: Science, Technology and the Clinical Art in Britain 1850–1914', *Journal of Contemporary History*, 20, 1985, pp. 503–20.
—— and Richard Dixey, 'Practising on Principle: Joseph Lister and the Germ Theory of Disease', in C. Lawrence (ed.), *Medical Theory, Surgical Practice*, Routledge, London and New York, 1992.
Lewis, Jane, *Women and Social Action in Victorian and Edwardian England*, Edward Elgar, Aldershot, 1991.
Lloyd, Genevieve, *The Man of Reason*, Methuen, London, 1984.
Lupton, Deborah, *Medicine as Culture: Illness, Disease and the Body in Western Societies*, Sage, London, 1994.
—— *The Imperative of Health: Public Health and the Regulated Body*, Sage, London, 1995.
Moore, J., *A Zeal for Responsibility: The Struggle for Professional Nursing in Victorian England, 1868–1883*, University of Georgia Press, Athens and London, 1988.
Morantz-Sanchez, Regina, 'Feminist Theory and Historical Practice: Rereading Elizabeth Blackwell' in Ann-Louise Shapiro (ed.), *Feminists Revision History*, Rutgers University Press, New Brunswick, 1994.
—— *Sympathy and Science: Women Physicians in American Medicine*, Oxford University Press, New York and London, 1985.
Moscucci, Ornella, *The Science of Woman: Gynaecology and Gender in England, 1800–1929*, Cambridge University Press, Cambridge, 1990.
Newman, Louise M., 'Critical Theory and the History of Women: What's at Stake in Deconstructing Women's History', *Journal of Women's History*, 2, 1991, pp. 58–68.
Pateman, Carole, *The Disorder of Women*, Polity Press, Cambridge, 1989.
Pelling, Margaret, *Cholera, Fever, and English Medicine, 1825–65*, Oxford University Press, Oxford, 1978.
Poovey, Mary, *Uneven Developments: The Ideological Work of Gender in Mid-Victorian England*, University of Chicago Press, Chicago, 1988.
—— *Making a Social Body: British Cultural Formation, 1830–1864*, University of Chicago Press, Chicago and London, 1995.
Pringle, Rosemary and Collings, Susan, 'Women and Butchery: Some Cultural Taboos', *Australian Feminist Studies*, 17, 1993, pp. 29–45.
Prochaska, F. K., *Women and Philanthropy in Nineteenth-Century England*, Oxford University Press, Oxford, 1980.
Rabinow, P. (ed.), *The Foucault Reader*, Penguin, Harmondsworth, 1991.
Rendall, Jane, *The Origins of Modern Feminism*, Macmillan, Basingstoke and London, 1985.
Rosenberg, C. E. (ed.), *Healing and History: Essays for George Rosen*, Science History Publications, New York, 1979.
—— 'Florence Nightingale on Contagion: The Hospital as Moral Universe' in his *Healing and History: Essays for George Rosen*, Science History Publications, 1979.

—— 'The Therapeutic Revolution: Medicine, Meaning and Social Change in Nineteenth-Century America', *Perspectives in Biology and Medicine*, 20, 1977, pp. 485-506.

Rubenstein, Annette, 'Subtle Poison: The Puerperal Fever Controversy in Victorian Britain', *Historical Studies*, 20, 80, 1983, pp. 420-38.

Shortt, S. E. D., 'Physicians, Science and Status: Issues in the Professionalisation of Anglo-American Medicine in the Nineteenth Century', *Medical History*, 27, 1983, pp. 51-68.

Showalter, Elaine, *Sexual Anarchy: Gender and Culture in the 'Fin de Siecle'*, Virago, London, 1992.

Shuttleworth, Sally, 'Female Circulation: Medical Discourse and Popular Advertising in the Mid-Victorian Era' in Mary Jacobus, Evelyn Fox Keller and Sally Shuttleworth (eds), *Body/Politics: Women and the Discourses of Science*, Routledge, London and New York, 1990.

Smith, F.B., *Florence Nightingale: Reputation and Power*, Croom Helm, London and Canberra, 1982.

Stevenson, L., '"Science Down the Drain": On the Hostility of Certain Sanitarians to Animal Experimentation, Bacteriology and Immunology', *Bulletin of the History of Medicine*, 29, 1955, pp. 1-26.

Anne Summers, '"Pride and Prejudice": Ladies and Nurses in the Crimean War', *History Workshop Journal*, 16, 1983, pp. 33-56.

—— *Angels and Citizens: British Women As Military Nurses 1854-1914*, Routledge & Kegan Paul, London and New York, 1988.

—— 'The Mysterious Demise of Sarah Gamp: The Domiciliary Nurse and her Detractors, 1830-1860', *Victorian Studies*, 32, 1989, pp. 365-86.

Tomes, Nancy, 'The Private Side of Public Health: Sanitary Science, Domestic Hygiene, and the Germ Theory, 1870-1900', *Bulletin of the History of Medicine*, 64, 1990, pp. 509-39.

Turner, Bryan S., *For Weber: Essays on the Sociology of Fate*, Routledge & Kegan Paul, Boston and London, 1981.

—— *The Body and Society: Explorations in Social Theory*, Basil Blackwell, Oxford, 1984.

—— 'The Practices of Rationality: Michel Foucault, Medical History and Sociological Theory' in Richard Fardon (ed.), *Power and Knowledge: Anthropological and Sociological Approaches*, Scottish Academic Press, Edinburgh, 1985.

—— *Medical Power and Social Knowledge*, Sage, London and Beverly Hills, 1987.

—— 'The Rationalization of the Body: Reflections on Modernity and Discipline' in Scott Lash and Sam Whimster (eds), *Max Weber, Rationality and Modernity*, Allen & Unwin, London, Boston, Sydney, 1987.

—— *Regulating Bodies: Essays in Medical Sociology*, Routledge, London and New York, 1992.

—— 'Recent Theoretical Developments in the Sociology of the Body', *Australian Cultural History*, 13, 1994, pp. 13-29.

Vicinus, Martha, *Independent Women: Work and Community for Single Women, 1850-1920*, Virago, London, 1985.

Walkowitz, Judith, *City of Dreadful Delight: Narratives of Sexual Danger in Late Victorian London*, University of Chicago Press, Chicago, 1992.

Wear, Andrew (ed.), *Medicine in Society: Historical Essays*, Cambridge University Press, Cambridge, 1992.

Williams, Perry, 'Religion, respectability and the origins of the modern nurse' in Roger French and Andrew Wear (eds), *British Medicine in an Age of Reform*, Routledge, London and New York, 1991.

——'The Laws of Health: Women, Medicine and Sanitary Reform, 1850–1890' in Marina Benjamin (ed.), *Science and Sensibility: Gender and Scientific Enquiry, 1780–1945* Basil Blackwell, Oxford, 1991.

Wohl, Anthony S., *Endangered Lives: Public Health in Victorian Britain*, J. M. Dent, London, 1983.

Wright, Peter and Treacher, Andrew (eds), *The Problem of Medical Knowledge: Examining the Social Construction of Medicine*, Edinburgh University Press, Edinburgh, 1982.

Youngson, A. J., *The Scientific Revolution in Victorian Medicine*, Croom Helm, London, 1979.

Index

abject, corpses 124–5
accoucheurs xiii, 63, 160 n 1
  hand washing 76
  hygiene 82, 124–6
  infection by 69, 81
  moral objections to 69–70
  status 65
  touch 72
  see also midwives
air, contamination 5–6, 137–8, 139
anatomy 111, 112
Anatomy Act 1832 110
Anderson, Elizabeth Garrett *see* Garrett, Elizabeth
animals, experimentation 121–2
anti-vivisection 121–2
antisepsis 128–32, 136–8, 141, 142
apothecaries 65
Ariès, Philippe 113
Armstrong, David 18, 78–9, 81
army, paradigmatic regime of discipline 44–5
asepsis xvi–xvii, 128–32, 136, 138–9
  surgeons 140–7

bacilli 139
bacteria 139
bacteriology 129, 130, 131
Baines, Mary Anne 12, 27
Barker, Mary 32
Barnes, Robert 76–7, 78
Barry, James 85, 94, 105
Beck, Snow 77
Bell, Joseph 58
bible-women 14
births, statistics 25
Blackwell, Elizabeth 85, 87–8, 93, 94, 95, 96, 98, 103–4
Bland-Sutton, John 112
blisters 133
blood, menstrual 37
blood poisoning 6
Bodichon, Barbara 88

bodies
  female *see* female bodies
  hygiene 16–19
  male *see* male bodies
  sexuality xii–xiii
bodily repulsiveness, nursing tasks 54
body snatching 109–10
Bolton, Gail 25
Bronfen, Elisabeth 114, 122, 123, 125
butchery 120
Butler, Josephine 96, 97, 122

carbolic 128–9, 130–1, 137, 138
Carter, K. Codell 75
castration 121
Chadwick, Edwin 3–4, 6, 7, 17
chlorine, disinfection 75, 76, 81
Churchill, Fleetwood 81–2
class relations, men and women 15
cleanliness
  doctors 141
  midwives 79, 145–6
  nurses 31, 76, 128, 136–9, 140–1
  surgeons 130–1, 140–7
Cobbe, Frances Power 96, 101–2, 103, 122
College of Physicians 65
College of Surgeons 65
Collings, Susan 120, 121
contagion, puerperal fever as 73–7
contagionist theory 6–7, 74–6, 129
Contagious Diseases Acts 97, 122
cordon sanitaire 78, 79
corpses 107
  sexualised 114
  trade in 109–10
  see also dead bodies
counter-irritation 133
Crimean War 45, 52, 91
cupping 133

Da Costa, J.C. 130

Davidoff, Leonore 36
Davies, Emily 86, 87, 112
deaconesses, nursing 22–3
Deaconesses' Institute, Kaiserwerth 45
dead bodies 107
  changing image 111–12
  cultural ambivalence 122–3
  see also corpses
Dean, Mitchell 25
deaths, statistics 25
Delhi Medical Mission 88–9
diagnosis, nurses 134–5
diphtheria 136
dirt
  as disorder 34
  matter out of place 19
discipline, nurses xv–xvi, 44–8, 53–4, 60
disease
  germ theory 2, 127, 128, 129–30, 131, 132, 137, 139, 141
  miasma theory 2, 5–7, 127, 129, 130, 131
  theories 3–8
disinfectants 128, 130–1, 139
dissection xiv
  by women 107–9, 112–13, 114–16, 125–6
  cultural attitudes 109–13, 123
  sexualising 113–22
dissection rooms 112
district nurses 27
Dixey, Richard 131
doctors xii
  and nursing reform 24
  as carriers of disease 64, 68–9, 79–80, 128
  cleanliness 141
  contagionist theory 6–7, 74–6, 81–2
  epitomising science and rationality 60–1
  masculinity xiii, xvi, 60–1
  nurses' behaviour towards 46–7
  obedience to 56
  professionalisation 64–5
  women xiv–xv, 72–3, 85–6, 87–94

Douglas, Mary 19, 34–5, 72, 123, 127
Dowling, Helena Pauline 92
Drysedale, Charles 92, 120
dying on the parish 110

Edmunds, James 66, 68–9, 70, 72–3
education, working-class girls 14
Elsthain, Jean Bethke 45
epidemic, puerperal fever as 73–7
erotic tendencies 59

Faithfull, Emily 85, 90, 91, 120
Fawcett, Millicent Garrett 68
female bodies 77–8, 111
  purity and pollution 35–9, 117–20
Female College of Midwifery *see* Female Medical College
female corpses 123
  sexualised 114
Female Medical College 66–9, 92
Female Medical Society 66, 68, 85, 88, 90, 100
femininity
  sexuality 57
  women doctors 86, 90–1, 94–105
feminism, Victorian xiv–xv, 95–6
Foucault, Michel xv–xvi, 4, 25, 41, 44, 46, 47, 123
Fox, Nicholas J. 140–1
Fry, Elizabeth 23, 51

Gamp, Sairey xv, 21, 30, 147
Garrett, Elizabeth 34, 66, 67, 68, 85, 86, 87–8, 94, 97–8, 100, 112, 121
gaze, anatomists 115
genitalia, male corpses 115
germ theory 2, 7, 75, 127, 128, 129–30, 131, 132, 137, 139, 141
Gordon, Alexander 145
Gregory, Samuel 90
Grosz, Elizabeth 39, 117, 119, 124
Guy's Hospital 26

hagiography, Nightingale 52–3
Hall, Catherine 36
hand washing 75, 76, 142, 144

hanging 109
head nurses 22
health
  as balance and equilibrium 132–3
  working-class 25
Holton, Sandra 98, 132, 137
home
  as centre for sanitary reform 1
  hygiene 16–19
  scientific view of 10
hospitals
  compared with homes xi
  design 6
  domestic conceptualisation of 10
  fear of 110
  hygiene 16–19
  military 93
  sanitary reform 32–3
  voluntary 25
housewives 11
humoral theory 140–1
Hygeia 7–8
hygiene 11, 139
  bodily and domestic 16–19
  working-class 25
  zones of 78–9, 80

India 87–9, 97
Indian Army 45
infirmaries
  paupers 15
  workhouses 25
International Congress on Charities, Correction and Philanthropy 44

Jack the Ripper 122
Jacobi, Mary Putnam 95
Jameson, Anna 16, 26, 50
Jex-Blake, Sophia 67, 100
Jones, Agnes 26, 52
Jordanova, Ludmilla 108, 114, 115, 126

King's College Hospital 23, 26, 68, 74, 80, 93
Koch, Robert 131
Kristeva, Julia 109, 123–4

Ladies' Health Society of Manchester and Salford 15
Ladies' Medical College *see* Female Medical College
Ladies' National Association for the Diffusion of Sanitary Knowledge 12
Ladies' Sanitary Association 11, 12–14, 27
lady-nurses 23
lady-probationers 23
Lawrence, Christopher 70, 71, 131
Lewis, Jane 15
Lister, Joseph 128, 129, 131, 137, 138, 145
Liverpool Workhouse Infirmary 26
London School of Medicine for Women 66–7
Lückes, Eva 34, 54, 133, 137–8
Lupton, Deborah 37

McBurney, Charles 129
male bodies xvi, 111, 115
  treatment by women doctors 92–3
male corpses 116–17, 123
male nurses 33
Martindale, Louisa 104
masculinity
  dissection rooms 112
  medicine 94–105
  surgery 120
masks, surgical 144–5
maternity wards, separation 79
Max, Gabriel 115
medical education 65
  dissection in 109–10
  women 66–9, 87–8
medical history, feminist xiii–xiv
medical knowledge, changing 24
medical missions 88–9
medical students 107, 109
  disorder 80
  dissection 110–11
  female 126
medicine
  feminine and masculine 94–105
  gendered xiv–xv

medicine (*contd*)
  masculinity 94–105
  quasi-religion 41–2
  reductionist 100–1
  scientific 132–3
  women in 65–7, 94–105
men, dissection by 107, 109
menstruation 37
miasma theory 2, 5–7, 31, 33–4, 127, 129, 130, 131
microbes 129–30, 139
microbiology 129, 131, 139
middle-class values, imposition 2, 32
middle-class women
  doctors 85–6, 92
  managerial role 26–7
  paid employment 50
  purity 36–7
Middlesex Hospital 87
midwives xiii, 28, 31, 88, 104
  cleanliness 79, 145–6
  contagion by 76–7
  discipline 80
  male 63, 65, 160 n 1
  professionalisation 67–8
  status 65
  training 98
  see also accoucheurs
military hospitals 93
military nurses 45
Mill, John Stuart 68, 97
modernisation 41
Montessori, Maria 124
morality
  and public health 3–5
  and sanitary reform 12
  responsibility for moral education 1
Morantz-Sanchez, Regina 95
Moscucci, Ornella 65
Munro, A. 135

National Association for the Promotion of Social Science 12, 14, 19, 25
National Health Society 104
Nature, healer 98
necrophilia 113

New Hospital for Women 97
new nurses xiii, xv, 21–2
  cleanliness 68–9
  conceptualised 28–34
  discipline 48, 80
  expertise 25–7
  purity 36–7
Newman, Louise xiv
Nightingale, Florence xv, 15, 21, 26, 31, 45, 89–90, 101, 102
  as statistician 25
  bodily hygiene 17–18
  Crimean War 91
  feminine medicine 96
  hagiography 52–3
  miasma theory 5–6, 7
  Notes on Nursing 10
  opposition to female doctors 68, 88, 97
  views on medicine 97–9, 132
Nightingale Fund 22, 23, 47, 53, 87
Nightingale Training School 93
nuns 57, 58
nurses xii, 104
  cleanliness 31, 68–9, 76, 128, 136–9, 140, 141
  discipline xv–xvi, 44–8
  femininity xiii, 35–6
  male 93
  old and new *see* new nurses; old nurses
  sexuality 56–61
  training 44–8, 98
  uniforms 58–9, 159 n 65
  women doctors and 87–94
  working class women 22–4
nursing
  nineteenth-century knowledge and practice 132–40
  reform 22–8
  religion and 22–3, 48–53
  secularisation 43
Nursing Record 28–9

obedience, nurses 44–5, 56
observation, nurses 134–5
Obstetrical College for Women *see* Female Medical College

Index

Obstetrical Society of London 65, 66
obstetricians 66, 160 n 1
obstetrics 71–2, 145
old nurses xv, 21–2, 68–9
 conceptualised 28–34
Osburn, Lucy 31–2, 52–3, 88–9, 93

Paget, James 138
Parkes, Bessie Rayner 11, 31
Pasteur, Louis 131, 145
patients
 male, contact with 59
 observation 47–8
pauperism 3
Pelling, Margaret 74
personal hygiene 81
physical work, nursing 53–4
physicians, status 65
physiology 112
Playfair, W. 130
pollution, women xi–xii, 36
Poor Law 3
Poor Law Amendment Act 1834 15
Poovey, Mary 3, 9, 15, 45, 52, 57, 59
Powers, Susan 14, 16
Pringle, Rosemary 120, 121
probationers 23
professionalism, nurses 56
prostitutes 36, 58
public health
 and theories of disease 3–8
 new 78–9
 women's involvement 8–16
puerperal fever 64, 130, 145
 aetiology 73–8
 exogenous or endogenous 77–8
 management 79–83
 spreading 69, 70
punishment 109
purity, women xi–xii, 36
Pusey, Edward 23
pyaemia 6, 145

Quaker Institution for Nursing Sisters 23
quarantine 78, 79, 80

rape, accoucheurs 70–1
rationalisation, control of sexuality 57
religion 45
 and nursing 48–53
 and sanitary reform 12, 14
religious sisterhoods, nursing 22–3
respectability, culture of 42
Rich, Adrienne 64
Richardson, Benjamin Ward 1, 7–8, 11, 18, 19, 31, 99–100
Richardson, Ruth 109–10
Rosenberg, Charles xi, 7, 101, 133–4
Rumsey, H.W. 68

St John's House Sisterhood 23, 26
St Thomas's Hospital 87, 88
sanitarian discourse xi–xii, 14, 127–8, 132–3, 136, 137, 139, 141
sanitary, usage of term 38
sanitary home missionaries 14
sanitary reform xv, 100, 104
 cleanliness and xi–xii
 new nurses and 30
 physical and moral issues 3–5, 35–6
 women's role 1–2, 11–16
Scharlieb, Mary 90, 104
schoolmistresses 14
scientific motherhood 12
secularisation 41
 nursing 43
self-sacrifice, nursing 51–2
Semmelweis, Ignaz 75–6, 145
senses, use of 134–5
separate spheres 95
sexuality 42
 bodies xii–xiii, 114
 dissection 113–22
 nurses 56–61
sick-room, environment 135
Simon, John xi, xii, 16–17
Sisterhood of Mercy 23
Sisterhood of the Holy Cross 23
sisters 22, 23
smells 6
statistics 25
sterilisation 129, 142

Summers, Anne 30, 45
surgeons
  asepsis 140–7
  potential polluters xvi–xvii, 69, 80
  status 65
surgery
  antiseptic practice 130–1
  by women 120–1
surgical gloves 144
Sydney Infirmary 31–2, 53, 88

teachers 14
therapeutics, nineteenth century 132–3
Thorne, Isabel 69
time, regulation, nurses 47
Tomes, Nancy 2
Tractarians 23
training, nurses 44–8
Turgenev, I.S. 114
Turner, Bryan 41, 43, 56–7, 60
Twining, Louisa 26
Tyler Smith, W. 71, 72, 76, 81
typhus 73

uniforms, nurses 58–9

venereal disease 36
ventilation 5–6
Vicinus, Martha 23, 45, 51, 53
vivisection 121–2

Walker, Mary 85, 94
Wardroper, Elizabeth 52
Weber, Max 43, 44

Wells, Spencer 121
Williams, Perry 2
Wohl, Anthony 5
women
  as objects of medicine xiv–xv
  dissection by 107–9, 112–13, 114–16, 125–6
  doctors xiv–xv, 72–3, 85–6, 87–94
  in medicine 65–7
  involvement in public health 8–16
  medical education 66–9
  middle-class *see* middle-class women
  moral authority 8–9
  moral reform 1
  purity and pollution xi–xii, 36
  sanitary reform 1–2
  surgery by 120–1
  working-class *see* working-class women
Wood, Catherine 135, 138
Wood, Christine 54
workhouses 15, 110
  infirmaries 25
working-class
  health and hygiene 25
  masculinity 33
working-class women 3
  domesticity 9–10
  nurses 22–4, 32

Yeo, Eileen 15
Youngson, A.J. 127

Zola, Émile 122